REAL WOMEN RUN

Real Women Run is an innovative feminist ethnography that consists of a series of linked essays and presentations about women who run at the intersections of queer, feminist, and running identities. Faulkner uses feminist grounded theory, poetic inquiry, and qualitative content analysis to examine women's embodied stories of running: how they run, how running fits into the context of their lives and relationships, how they enact or challenge cultural scripts of women's activities and normative running bodies, and what running means for their lives and identities. During a two-and-a-half-year ethnography with women who run, Faulkner engaged in an intersectional qualitative content analysis of websites and blogs targeted to women runners, a grounded theory poetic analysis of 41 interviews with women who run, and participant observation at road races.

Real Women Run speaks to the call for a more physical feminism. This ethnography sees women's physical and mental strength developed through running as a way to embrace the contradictions between a deconstructed focus on the mind/body split and the focus on individuals' actual material bodies and their everyday interactions with their bodies and through their bodies with the world around them.

Sandra L. Faulkner is Professor of Communication and Director of Women's, Gender and Sexuality Studies at Bowling Green State University, Ohio. Her research interests include qualitative methodology, poetic inquiry, and the relationships among culture, identities, and sexualities in close relationships. Faulkner is the recipient of the 2013 Knower Outstanding Article Award from the National Communication Association, and the 2016 Norman K. Denzin Qualitative Research Award.

INNOVATIVE ETHNOGRAPHIES

The purpose of this series is to use the new digital technology to capture a richer, more multidimensional view of social life than was otherwise done in the classic, print tradition of ethnography, while maintaining the traditional strengths of classic, ethnographic analysis.

Series Editor: Phillip Vannini, Royal Roads University

Available

Ferry Tales: Mobility, Place, and Time on Canada's West Coast by Phillip Vannini

Digital Drama: Teaching and Learning Art and Media in Tanzania by Paula Uimonen

Concrete and Dust: Mapping the Sexual Terrains of Los Angeles by Jeanine Marie Minge and Amber Lynn Zimmerman

Water in a Dry Land: Place Learning Through Art and Story by Margaret Somerville

My Father's Wars: Migration, Memory, and the Violence of a Century by Alisse Waterston

Off the Grid: Re-Assembling Domestic Life by Phillip Vannini and Jonathan Taggart

Schooled on Fat: What Teens Tell Us About Gender, Body Image, and Obesity by Nicole Taylor

REAL WOMEN RUN

Running as Feminist Embodiment

Sandra L. Faulkner

NEW YORK AND LONDON

First published 2018
by Routledge
711 Third Avenue, New York, NY 10017

and by Routledge
2 Park Square, Milton Park, Abingdon, Oxon, OX14 4RN

Routledge is an imprint of the Taylor & Francis Group, an informa business

© 2018 Taylor & Francis

The right of Sandra L. Faulkner to be identified as author of this work has been asserted by her in accordance with sections 77 and 78 of the Copyright, Designs and Patents Act 1988.

All rights reserved. No part of this book may be reprinted or reproduced or utilised in any form or by any electronic, mechanical, or other means, now known or hereafter invented, including photocopying and recording, or in any information storage or retrieval system, without permission in writing from the publishers.

Trademark notice: Product or corporate names may be trademarks or registered trademarks, and are used only for identification and explanation without intent to infringe.

Library of Congress Cataloging-in-Publication Data
A catalog record for this book has been requested

ISBN: 978-1-138-21829-1 (hbk)
ISBN: 978-1-138-21830-7 (pbk)
ISBN: 978-1-315-43785-9 (ebk)

Typeset in Bembo
by Out of House Publishing

This book is for all the (women) runners I have met, run with, and those I have yet to meet. We are real women running.

CONTENTS

Acknowledgments	*viii*
1 Women Running	1
2 Woman, Running	9
3 Real Women Run	59
4 Women Running Online	90
5 Running as Feminist Embodiment	105
Appendix	*122*
Index	*126*

ACKNOWLEDGMENTS

Part of the following articles were adapted for Chapter 2:

Faulkner, S. L. (2013). Notes from a pretty straight girl: Questioning identities in the field. *Liminalities, 9*(2), 39–48. Retrieved from http://liminalities.net/9-2/faulkner.pdf

Faulkner, S. L. (2016). Cancer triptych. *Health Communication, 31*(8), 1043–1046. doi:10.1080/10410236.2015.1020262

Faulkner, S. L. (2016). Postkarten aus Deutschland: A chapbook of ethnographic poetry. *Liminalities, 12*(1). Retrieved from http://liminalities.net/12-1/postkarten.html

Faulkner, S. L. (2016). TEN (The promise of arts-based, ethnographic, and narrative research in critical family communication research and praxis). *Journal of Family Communication, 16*(1), 9–15. doi:10.1080/15267431.2015.1111218

Faulkner, S. L. (in press). Bulls-eye: An intimate history of guns. *International Review of Qualitative Research.*

Dr. Dinah Tetteh and Dr. Erika Behrmann worked on an earlier version of Chapter 4. We presented this work at the 2016 Central States Communication Association annual meeting in Grand Rapids, MI. (Faulkner, S. L., Tetteh, D. A., & Behrmann, E. M. Women Who Run: Identity, Embodied Experience, and Gendered Expectations in Popular Women's Running Websites.)

Part of the poem 'Love & Guns: Orlando' (Faulkner, S. L., forthcoming, *Qualitative Inquiry)* appears in Chapter 5. The poem, *History of Body*, in Chapter 5 was a finalist in the Gulf Stream Magazine Summer Contest and appears in *Gulf Stream Magazine* (October, 2017).

1
WOMEN RUNNING

The runner begins anew every time she puts on her shoes. (Menzies-Pike, 2017, p. 84)

I run because that is who I am. (41 women runners)

Running and sports in general is transformational. When a woman feels her own strength, it's empowering. She believes in herself and knows she can do more. It changes everything. (Katherine Switzer quoted in Loudin, 2017, para. 14)

Women and Running

Women are running in record numbers, as evidenced by forums in popular online running sites such as *Women's Running* and the emergence of self-made blogs with stories of women who run like *Fit and Feminist* (https://fitandfeminist.com/about/) and *Fat Girl Running* (http://fatgirlrunning-fatrunner.blogspot.com/). In the past two decades alone, the number of recreational women runners who compete in racing has increased ten-fold. Women now outpace men who run races, at about 1.19 million female runners total (RunningUSA.org, 2015), and in 2014, 61% of half marathon racers were women (RunningUSA.org, 2014). In 2015, 9.7 million women finished road races nationwide with more women than men running the half marathon, 10K, and 5K distances (RunningUSA.org, 2016). On World Running Day in 2017, *Runner's World*, a popular running media source, announced that Betty Wong Ortiz would be the first woman to be editor in chief in the 51-year history of the magazine (Runner's World Editors, 2017).

2 Women Running

What accounts for this surge in women who run? One reason could be the passage of Title IX in 1972, which increased opportunities for girls to participate in sports after making it illegal to discriminate against girls and women in federally funded education. Women were not allowed to compete in Olympic Games and official distance races until 1972, when the women's 1,500 meter was added at the Olympics; in 1984 the women's marathon was added; in 1988 the women's 10,000 meter became an Olympic event; in 1996 the women's 5,000 meter was added; and in 2008, the women's steeple chase was added as a woman's Olympic event (Burfoot, 2016). This also may have made women's running more attractive to the average runner. Kathrine Switzer, the first women to be officially registered to run the Boston Marathon in 1967, argued that women run for physical and mental reasons, in an interview about her 50th anniversary run at the 2017 Boston Marathon:

> What's changed hugely is women have become empowered from running. I submit the reason women run at all is because of that sense of accomplishment, self-esteem and confidence running gives them. Once you start running, you question other things in your life that don't make you happy. (Pattillo, 2017, para. 8)

Switzer registered for the 1967 Boston Marathon under the name K. Switzer and ran the race with the number 261. The photo of the race director, Jock Semple, trying to yank her off of the course became iconic of women and running.

Shalane Flanagan, who won the silver medal in the 2008 Beijing Olympic 10,000-meter race and won the 2017 New York City Marathon, credits women who came before her—Kathrine Switzer, Grete Waitz, Lorraine Moller, Cheryl Flanagan (her mother!), Joan Benoit Samuelson—for making running easier for her and all women.

> And they didn't just do it for Olympians like me. They did it for all women runners. Because one of the things they learned is that running improves your self-confidence and can give you a sense of possibility that carries over to all parts of your life. (Burfoot, 2016, p. xi)

Others credit the popularity of running for women to Oprah Winfrey, who trained for and ran the 1994 Marine Corps Marathon.

> Oprah's marathon finish had a galvanizing effect on millions of women runners around the world. Joan Benoit had won the Olympic Marathon a decade earlier, but her great victory produced few women runners … It was much easier to relate to an Oprah than to an Olympic gold medalist. "If Oprah can do it, so can I." That was the rallying cry that launched the huge boom in women's running that began immediately after Oprah's finish. (Burfoot, 2016, p. 240)

Even the marketplace has shifted this once male-dominated field to a focus on products and races aimed at women runners. For example, the *Disney Princess Half Marathon and Tinker Bell Half Marathon Weekends* (www.rundisney.com/princess-half-marathon/) where women and girls are promised "a storybook weekend chock full of magical runs, spectacular surprises, (and) amazing sights," the *Divas Half Marathon and 5K series* (www.runlikeadiva.com/), with the slogan "Run Like A Diva®," and the *Run Like A Mother* 5K series (www.runlikeamother.com/) urging women to "Run Like a Mother®," validate the explosion of women's participation in running in the past two decades. The Oakley New York Mini 10K (www.nyrr.org/races-and-events/2017/nyrr-new-york-mini-10k) lays claim to being the original race just for women.

> As the world's first women-only road race, the Oakley New York Mini 10K began with just 78 women in 1972, the same year Title IX became law and the Olympics introduced 1,500 meters as its longest distance for female athletes. The Mini, named after the miniskirt, now attracts a sold-out field of 5,600 women who run for personal bests through Central Park alongside Olympians like Paula Radcliffe, Deena Kastor, and Kim Smith, who are vying for the race's $34,950 prize purse. (Bruning, 2017)

In addition to women-only races, there are shirts, bags, and other products emblazoned with sayings such as "I Run like a Girl. Try and Keep Up," and "Forget the Glass Slippers. This Princess Wears Running Shoes." Women can buy running shoes and anti-chafing body glide in pink, made just for women, too. The marketing of running with pink power is in full bloom.

Online running groups and organizations like Katherine Switer's *261 Fearless®* (www.261fearless.org/) support women's participation in running through the use of communication, running support, and community:

> Pronounced TWO-SIX-ONE Fearless, we are a global supportive community which empowers women to connect and take control of their lives through the freedom gained by running. Through a series of non-competitive running clubs and private communication channels, we provide networking, healthy running support and education, and a sisterhood to women all over the world. (para. 1)

Other organizations view running for women as more than just a leisure activity. The "Run Like a Mother®" series proclaims that running is a lifestyle:

"It's not just a race, it's our way of life."
Run Like a Mother

Run Like a Mother® fuels a woman's journey toward health and wellness. It starts with our Run Like a Mother 5K! (www.runlikeamother.com/)

4 Women Running

Of course, there are local running groups connected to specialty running stores or groups like *Wear Blue* and *Frontrunners* (see Chapter 3), which women may join to support important causes and be part of a particular running community in their locale.

Real Women Run

This ethnography I call *Real Women Run*, consists of a series of linked essays and presentations at the intersection of women runners' stories, feminism, identities, and running in everyday relational life. I investigate how women's narratives of running subvert mainstream discourses of what being female and being active mean in terms of identity, motivation, and practice. I was a participant observer in running events at the 2014 Gay Games in Cleveland/Akron, OH, USA (www. gg9cle.com) and ran the 5K, 10K, and half marathon road races and 13 other road races in 2014–2017, such as the Flying Pig Half Marathon in Cincinnati, OH and the Munich Marathon 10K.

In addition, I interviewed 41 women across the US who run and talked with them about their experiences of running: how running makes them feel, why they run, what running means to them, how they run, and running and their social networks. I present women's running stories through a poetic analysis of the interviews (Faulkner, 2009, 2017), my own experiences of running, including my participant observation at the Gay Games and other road races in an auto-ethnographic memoir, and a critical content analysis of websites, running organizations, and blogs targeted toward women runners. I focus on representing women's lived embodied experiences.

I am interested in women's embodied stories of running: how they run, how running fits into the context of their lives and relationships, how they enact or challenge cultural scripts of women's activities and normative running bodies, and what running means for their lives and identities. Through ethnographic work including interviews, poetic inquiry, participant observation, and textual analysis of women's writing about running, I add to the literature on how cultural and relational processes such as the enactment of values, attitudes, and identities are embodied practices (Faulkner & Hecht, 2011). In short, I argue that women's embodied experiences matter. I wrote this ethnography on women and running using a feminist lens, because "not many books about running speak to women, and when they do, it's often about weight loss. As for feminist analyses of running, they were drowned out by exhortations to 'run like a girl' (Menzies-Pike, 2017, p. 3).

Reasons Women Run

Women begin running for a variety of reasons, such as for exercise, to lose weight, and because of encouragement from family and friends; women keep running to stay in shape, to be healthy, and to relieve stress (Running USA, 2014). When I talked with women about running, they told me they experienced running as social and solitary, pleasurable and painful, dangerous and empowering. *Real Women Run* is about women running; about identities in motion, the inseparable mind–body connection, and running as solitude, physical and emotional strength, and community. Their stories of running are detailed in Chapter 3. Talking and writing about running from a feminist perspective helped me to see the importance of embodiment:

> The language of running is gendered too. As I ran more, I began to listen closely to the inflections that distinguish cautionary details about women runners from heroic epics about the men who run after them. Subjects versus objects: one runner might experience the thrill of the chase; the other, the terror of being chased. (Menzies-Pike, 2017, p. 6)

Running and Embodied Experience

> But here's the thing—not all of us who run do so because we are trying to look like Kara Goucher. We do it even though we may not be thin or slim-hipped or flat-chested. And not all of us are out there forcing ourselves through the miles because we secretly hate ourselves. We run because we love it. For most of us, running has made our lives better. I know this is true for me. Running has made me more confident, braver, tougher. I suspect that if you go to a road race on any given Saturday morning and ask the women standing around, most of them would tell you the same. (Constantine, 2012)

Perhaps most important are the socio-cultural influences on women who run, how women's running bodies are embedded in larger cultural discourse about appropriate ways of being (Jutel, 2009). The overwhelming cultural image of a woman runner and the normative running body is that of elite runner Kara Goucher—white, thin, straight, fast, feminine, middle-class and disciplined (Hanold, 2010). We need to acknowledge the bodies and stories of other women and their reasons for running. This ethnography contributes to an embodied feminist project. Embodiment for women who run means self-determination, their choice of identities, and an integration of health and running, even if women must react against the dominant image of a female running body and skirt the dialectic between safety and danger, empowerment and marketing, relationships and solitude (see Chapter 4). Few studies examine women's experiences of running via their perceptions and bodily experience. Allen-Collinson (2010) argues for a phenomenological perspective, which is "grounded in the carnal, 'fleshy,' lived, richly textured realities of the moving, sweating, sensuous female sporting body, which of course also holds cultural meanings, significances, purposes and interests" (p. 293). Through memoir-writing about feminist identity and running (see Chapter 2) and poetic analysis

6 Women Running

of my interviews with women who run (see Chapter 3), I argue that running offers women a sense of control in the everyday moments of difficulty related to work, relationships, and health. I work with women's stories to situate the meaning of running, why they run, and how running functions in their lives from their embodied perspectives (see Chapter 5).

Running Ahead

> I'm finally doing the research project I've thought about for years … I'm running toward the feminist ethnographer I want to be. (Faulkner, Chapter 2, this volume)

Real Women Run is a feminist embodied ethnography that defies the mind–body split because of the attention paid to the material and the discursive; this work takes up emotional, physical, and ideological space. My running helped my writing, and my writing helped my running. I used poetic inquiry as a feminist ethnographer to contend with dominant discourses about women's running bodies, identities, and gendered/sexualized roles and the concomitant evaluations by acknowledging, examining, and altering the complex reality that not all women who run are young, long, and lean (Faulkner, 2009, 2017). There are fat runners, and old runners, and average runners who sometimes walk during races. Running is a way to build physical and emotional strength to challenge and resist normative running bodies, typical femininity, and staid expectations. We see how women run toward and run away from expectations, relationships, and ways of being.

In Chapter 2, Woman, Running, I present an autoethnography on women and running using creative nonfiction to show my embodied experiences of running while female. I depict my emerging feminist consciousness from grade school to the present through the details of interaction in close relationships and the connection between relational processes and running practice. I use my poetic analysis of this ethnography on women and running, which includes participant observation at the 2014 Gay Games to tell a story of woman as runner, woman as feminist runner, woman as feminist partner runner, and woman as feminist mother runner. Scenes of running in everyday contexts, in road races, with friends, and of not running because of physical and psychic injuries, are interspersed with the use of haiku as running logs to make an aesthetic argument for running as feminist and relational practice.

Chapter 3, Real Women Run, highlights the voices of 41 women runners I interviewed about their embodied experiences running and being a runner. In an essay woven with scholarly research, poetic transcription, and women's reflections, I answer the question, *What are women's embodied experiences of running and being a runner?* Running for the women I talked with was about being in control, being blessed, being healthy, setting goals and meeting them, challenging the self and body, being strong, being safe, and, for some, part of a spiritual practice. Running was part of their identities and a way of life. Their perceptions of their running bodies, what running means to them, and how running fits into their lives and social networks is shown through the themes: *running as health*; *running as accountability*; *running as relational practice*; *running as safety and danger*; *running for a reason*; *running with your body*; *running as expansion*; and *running as self-definition*.

Chapter 4, Women Running Online, focuses on a critical content analysis of the gendered expectations in the popular women's running websites, *Zelle* of *Runner's World* and *Women's Running*, and women-focused running blogs. I discuss my interviews with women runners and select content from women's running blogs—*Run Like a Girl, 261 Fearless®, Run Like a Mother®, Fit and Feminist, Fat Girl Running*—to show real women's experiences running with their mother, fat, queer, and average bodies as a counterpoint to commercially produced material. The online narratives women runners present subvert the dominant cultural image of the white, thin, straight runner who runs solely to maintain aesthetics. These counter narratives demonstrate how women run against idealized images of a "running body."

In Chapter 5, Running as Feminist Embodiment, I articulate the connection between running and writing. My running practice is tied to my writing practice, so it was impossible to talk about my ethnographic analysis of women and running without talking about running *and* writing. I used poetic inquiry as an analysis technique and discuss how this was the key to organizing and demonstrating this project as a feminist embodied ethnography; poetic inquiry as analysis, as representation, and as embodiment. I argue how the haiku as running log that I weave into my running narrative in Chapter 2 is a form of embodied inquiry, and how women's narratives about running demonstrate the idea of physical feminism.

The web material (see http://innovativeethnographies.net/realwomenrun) that accompanies the written work shows the embodied fieldwork through sound and image (Harris, 2016). Using video as a component in this ethnography is a way to help you think differently about women and running; it is not just a digital version of *Real Women Run*, but another nuanced layer of women's embodied experiences of running. In addition, the soundscapes help present my embodied presence in this ethnographic project (Makagon & Neumann, 2009). The sounds of running—the noise, the grunts, the breathing, the encouragement, the disappointment—jog you through training runs, races, and the in situ embodiment of a sport real women enjoy and loathe.

I hope you pace with us through the words, images, and sounds of women jogging, crying, raging, laughing, sprinting, and walking through their running lives.

References

Allen-Collinson, J. (2010). Running embodiment, power and vulnerability: Notes towards a feminist phenomenology of female running. In E. Kennedy & P. Markula (Eds.), *Women and exercise: The body, health and consumerism* (pp. 280–298). London: Routledge.

8 Women Running

Bruning, K. (2017). The top 10 women's races: Where running like a girl is rewarded with wine, flowers, jewelry, and other ladies'-only prizes. *SHAPE*. Retrieved from www.shape.com/fitness/cardio/top-10-womens-races

Burfoot, A. (2016). *First ladies of running: 22 inspiring profiles of the rebels, rule breakers, and visionaries who changed the sport forever.* New York, NY: Rodale.

Constantine, C. (2012). Should women run? You're damn right. They should. *Jezebel.* Retrieved from http://jezebel.com/5947053/should-women-run-youre-damn-right-they-should

Faulkner, S. L. (2009). *Poetry as method: Reporting research through verse.* New York, NY: Routledge.

Faulkner, S. L. (2017). Poetic inquiry: Poetry as/in/for social research. In P. Leavy (Ed.), *The handbook of arts-based research* (pp. 208–230). New York, NY: Guilford Press.

Faulkner, S. L., & Hecht, M. L. (2011). The negotiation of closetable identities: A narrative analysis of LGBTQ Jewish identity. *Journal of Social and Personal Relationships, 28*(6), 829–847. doi:10.1177/0265407510391338

Fit and Feminist (n.d.). Retrieved from http://fitandfeminist.wordpress.com

Hanold, M. (2010). Beyond the marathon: (De)Construction of female ultarunning bodies. *Sociology of Sports Journal, 27*, 160–177.

Harris, A. M. (2016). *Video as method.* Oxford: Oxford University Press.

Jutel, A. (2009). Running like a girl: Women's running books and the paradox of tradition. *Journal of Popular Culture, 42*(6), 1004–1022. doi:10.1111/j.1540-5931.2009.00719.x

Loudin, A. (2017, April 10). 50 years after sparking a revolution, an icon runs Boston again. *Runner's World.* Retrieved from www.runnersworld.com/boston-marathon/50-years-after-sparking-a-revolution-an-icon-runs-boston-again

Makagon, D., & Neumann, M. (2009). *Recording culture: Audio documentary and the ethnographic experience.* Thousand Oaks, CA: Sage.

Menzies-Pike, C. (2017). *The long run: A memoir of loss and life in motion.* New York, NY: Crown.

Pattillo, A. (2017, March 29). Kathrine Switzer on her return to the Boston Marathon. competitor.com. Retrieved from http://running.competitor.com/2017/03/news/kathrine-switzer-return-boston-marathon_162927

Runner's World Editors (2017, June 7). *Runner's World* gets its first female editor in chief: Betty Wong Ortiz has deep background in fitness and running. Retrieved from www.runnersworld.com/about-runners-world/runners-world-gets-its-first-female-editor-in-chief

Running USA (2014). *Annual half-marathon report.* Retrieved from www.runningusa.org/2014-half-marathon-report?returnToannual-reports

Running USA (2015). *Statistics.* Retrieved from www.runningusa.org/statistics

Running USA (2016). *2016 State of the sport: U.S. road race trends.* Retrieved from www.runningusa.org/state-of-sport-us-trends-2015

2

WOMAN, RUNNING

Middle-Aged Run

You know you are middle aged
when the mist at the top of the hill
on your run beside the brook
burping over lichen caressed rocks
makes you slip; lost footing
on the urban nature trail,
faux wood planks crank
your 20-year-old plantar fasciitis,
widen the leak in your pelvic floor
make you sure you are the blood mobile
all urine and sweat and smelly fluids.
You know where every porta-potty
sits and urinate in daylight
in front of the elementary school
where your daughter learns
because you won't unload
in your pants, down your ankles.
You know that it gets
better after the fourth mile,
but all you can think is this:
what a serial killer view,
if I stop to pee in front of a tree
who will be waiting?
The most dangerous thing to do
is run alone with all of this age. (Faulkner, 2014)

July 8, 2014: Woman, Runner

a woman runner
is a woman who runs is
a woman runner

July 7, 2016, Injured @ 372/1000 Pledged Miles, Age 44, Bowling Green, OH

My daughter and I walk to the public library to claim her 5K prize for 192 minutes of reading. She sprints ahead with her Jack and Annie book, *Summer of the Sea Serpent*, to return while I manage not to trip over the curb, my left arm in a pillow sling to support the shoulder with two stainless steel screws fastened into the os acromiale. "It was too large to remove and keep the shoulder stable," Dr. Gomez told me in the post-surgery haze, handing me two grayscale X-ray pictures of his work. Before the surgery, I told him that my husband, Josh, wanted the bone as a souvenir, so the photos were the next best things. I am pleased with the idea of a youth summer reading program with the slogan, *Reading Marathon: On Your Mark, Get Set, Read!*, given I cannot run for the foreseeable future. I am more than grumpy about it, because all I can do is read and write about running. With one hand. Art is not imitating life, in this case. The Faulkner curse of accident proneness strikes again. For example, during my surgery, my father was at home in suburban Atlanta using his pocketknife to enthusiastically open a late package. When I called my mother to let her know I was alive after my operation, she told me he was angry as he sliced into the box and "bled like a stuck pig" all over the kitchen table, floor, and countertops from the deep gash he sliced into his wrist. They called their neighbor Emmett, a retired EMT, to make sure he didn't hit a major artery. Emmett cleaned up the blood, while my mother bandaged the wrist Dad waited till morning to get stitched up. I imagine Emmett enjoyed a glass of Irish whisky with Dad at the freshly cleaned kitchen table like he usually does when he checks on my disabled parents.

While technically this extra bone is not an accident, but a genetic gift, it has benched me for at least 5 weeks. I look down and notice that the black cotton short-sleeved button-up-shirt—I was going to write blouse, but I gave up blouses in the eighth grade—is inside out. "Hey Mimi, look at the Mama's shirt!" I say as we swoosh through the automatic doors into the library. *Wow, I managed to do that with one arm.* I also managed to use the elliptical machine at the gym earlier. I didn't ask Dr. Gomez whether this was allowed. I learned to never ask questions I didn't want to know the answer to when my feminist consciousness began to emerge in my late teens. I became a feminist when I no longer cared what others thought about me and held the inflexible woman love at my core tighter than my runner hamstrings.

I have always been a runner, though my path looks like a traffic jam with all of the start-stops, pauses, and parentheticals that color in the jagged line graph of what it means to be a woman who runs, a woman runner. Claiming the title runner in my list of relational and intellectual identities is political when who counts as a runner and what it means to run are defined through a masculine lens. Running is my feminist act because when I run, I sweat out the feminine should: *girls are not good at math; shy stubborn girl nerds with knock-knees are not athletic; girls do not stink; women should want to be married; middle-aged female bodies should not run unless the pudgy parts are draped in folds of fabric.*

As I watch my daughter play a matching word game on the library computer, I wonder if I can count the 29 one-armed miles on the elliptical this July toward the 1,000 miles I publicly pledged to run on Facebook in January? That would make 401 miles in 2016, only 599 more to go. I know I should stop being competitive and focus on the healing process, but like most difficult things in life, I am impatient with the wait for the crappy parts to go faster, for the screws to fuse into my bones, for the 5 weeks off from running to be over. Mimi and I talk about how she can keep reading, and I will log in the minutes read in her electronic account, the thermometer of progress heating up toward the 1,572 minutes for the 26.2 miles of reading win. I encourage her hyper-focus. "You know, Mimi, Mama was slow to read. You are doing better than me at your age." This confession makes Mimi smile and begin another chapter in her new Jack and Annie book. I have continued to run through all of the inevitable limitations; my running maladies, both physical and social, have crafted my embodied self, tough and female.

I placed in races in elementary school when being a kid meant being in constant motion and running a first language. I had forgotten about the wins until I found the torn and bug-eaten red ribbons in a damp cardboard box rescued from my parents' basement—fifth-grade field-day competition: "2nd place mile run; 2nd place 5-lap run." I likely will never place in another race, because I'm a middle of the pack runner. I describe myself as slow, though I usually place in the upper third of whatever age group I happen to circle on a race form: 9 **X** 11 **X** 19 **X** 25 **X** 29 **X** 30 **X** 40 **X** 41 **X** 42 **X** 43 **X** 44 **X**. I remember thinking that adults were boring because they liked to sit and visit, meaning they sat and talked all morning and afternoon, slowness on purpose. And now I read on the *Women's Running* site about a Copenhagen heart study suggesting that slow joggers outpace moderate and over-the-top joggers: the slower you run, the longer you live.

Slow Run
hot, stuck. stop to pee
farmer blow to spit the miles
warm and nasty breath

1983, Mathis Dairy Fun Run, Age 11, Decatur, GA

I entered the 1-mile fun run because my physical education teacher encouraged me with the race flyer she handed me during recess. "You can run this race." She must have remembered my victorious field-day runs around the gravel playground

12 Woman, Running

at Murphy Candler, though I spent my free time with the nerds in the highest reading group after giving up months of recess in the third grade. Mrs. Elkins coached me out of shyness and slow reading at the back of the pack as I turtled through *Dick and Jane* while my peers played kick-ball. Being given encouragement for my body as athlete and runner was alien. I still considered myself a nerdy knock-kneed klutz who could trip over objects that didn't exist in the physical world. I am skilled in the art of accidents, and stitch merit badges onto my human Girl Scout sash—broken nose from smashing into a suitcase with my face after a summer camp bunk-bed incident; deep gash in right leg from broken glass in trash bag when taking it out to the curb.

My mom dropped me off at the start line and told me she would wait for me at the finish line. I can't remember if I felt nervous. I concentrated on not tripping as I wound up, down, and around the grassy pasture hills. The family nickname my younger brother gave me stuck in my mind on repeat. "Nurse Nerd. Nurse Nerd. Nurse Nerd!" Bill taunted me at the top of the basement stairs as I tripped up them, my adolescent legs like a puppy's gangly feet, always in my way. But in this race, I was a unicorn sprinting around Rainbow Drive, my legs a blur, my breath puffs of speed, a heat cloud of effort swirled the dirt behind me. I was Rainbow Dash with Pegasus wings flying over other kids. I ran so hard my nose started bleeding, but I didn't notice until I crossed the finish line. A race volunteer told me to go to the award ceremony in another part of the pasture as the blood dripped onto my race T-shirt. When I overheated, my nose opened up like the warm and fickle Georgia skies. My mom and I couldn't make it stop. I missed the award ceremony; hearing my name called out loud and holding up my third place medal. Mom told me there would be other races, but we didn't know that would be my last race for almost a decade because of the Faulkner proclivity for dramatic accidents.

I still have the medal, and I kept the bloody shirt for years, too, not wanting to wash off a shy and nerdy girl's proof of victory. I wore it out, needing to tell the story of the splattered spots of rust to whoever would ask. Most people, though, commented about some grooming foul I had committed: "You have dirt on your shirt, gross."

"That isn't dirt. It's blood!" I replied. Most likely I threw in a smirk for maximum effect.

July 10, 2014: Hard Run
loop the hospital
just in case you need to drop
tired bird of prey

1984, Recess, Age 12, Murphy Candler Elementary, Lithonia, GA

One spring in the seventh grade, I was outside during recess practicing to run hurdles—ropes stretched across lunchroom chairs—for seventh-grade field day and what my PE teacher and I hoped would be the start of my track career. I knew that the rigged hurdles were not a great idea, especially when tubby Tommy Asbury taunted me from one side of the field, tossing gravel and stupid words at my feet. I ignored him and concentrated on getting my knees up and clearing the ropes, feeling the swoosh of air under my legs. Then he kicked one of the chairs in my path. I caught my legs in the rope and tumbled to the ground in an ungraceful heap. *What a mean stupid boy.* Now I think this was some kind of elementary flirting in a world where violence against women is the norm. This may mark the first check in the bank of my emerging feminist consciousness. Tommy Asbury got detention, and my sprained ankle kept me from competing in field day. That was my last run for years, my trophy a right ankle with a fat pocket of scar tissue that sprains if I walk the wrong way. I skipped track and became a high school band nerd.

July 14, 2014: Sick Run
didn't you just pass
that sign and that sign and that
mucous dizzy fuzz

Spring 1990, Feminist in Training, Age 18, Evansville, IN

I rediscover running my first semester in college because of my crush on Tim, the Catholic fraternity boy I met at the university radio station. Tim ran track in high school, while I played the bass clarinet or tenor or baritone saxophone in every band from marching to concert to jazz. I lace up and run my band-nerd self around the rectangular perimeter of the university—Walnut Street in front of my co-ed dorm past the bus-stop shelter Tim and I would nestle in for private conversations; left onto Rotherwood Avenue past the duplex I would escape the dorm for the following year; another left to Lincoln Avenue by the Shell station where in another 2 years I would trade couch cushion change and my pink lungs for generic smokes; left onto Weinbach Avenue near the drug-store where my future roommate would write bad checks for liquor, and back to Walnut. I ran once or twice a week, willing the romance to keep pace.

And the shoes matter. I think of how and when I will need to run before I go to England to study next year as I huff around the university. I want to get in shape and become a runner. I always wear running shoes to my Friday night shift at the university radio station that begins at 11 pm and ends at 3 am. The callers to *Alternative Wavelengths on WUEV*, many of whom I hang up on when they drunkenly yell at me to play Danzig, make the early morning quiet on campus not a welcome sound of solitude. I know the moral of the story: girl alone in the dark always means blame the female. Earlier this year, I walked alone to a house party by the river, because I couldn't find a friend with a car to drive me. When I crossed the tracks over I-44, I sensed that I was being followed. My peripheral vision is not

14 Woman, Running

terrific, so I switched to the sidewalk on the other side of the street. My disturbing shadow switched sides. I heard the sounds of another pair of shoes marking the pavement, an echo that began in my ears and traveled through the blood vessels in my shoulders and arms to my chest and filled my legs with the energy to sprint two or three or four blocks to the party. I got a ride back to campus and a new criterion for choosing everyday footwear: *Can I run in these?*

I stopped running consistently the next fall when Tim dumped me in the second floor hallway of his frat house, brothers streaming by on the way to the showers. "I don't want to see you anymore. I never loved you," he said. The unilateral break-up line deterred any response but walking out, then running back to my duplex on Rotherwood so that the tears would fly off my face like invisible wings.

2015: Fall Run
colors run the sky
muscles burn like dropping leaves
fall into a run

1991, The Naked Mile, Age 19, Harlaxton College, England

I met Stephanie spring of my sophomore year spring during the requisite study abroad semester at Harlaxton College in England, 7 months after the break-up bomb. She was an army kid with European sensibilities and feminist wit. I mailed my mom a postcard with a picture of a woman ironing a shirt crossed out with an X, the F word—FEMINIST—pasted prominently on the front. I came out as a feminist in pen on the back, telling her it was because of her good influence. I had to endure my younger brother's singsong taunt "feminist, feminist, feminist" when I returned home in the summer.

We planned to run the traditional naked mile down the gravel driveway from the Gothic Harlaxton Manor one crepuscular evening in April. "Would it still count if I wore running shoes?" I asked. My feet always hurt.

"If you don't wear socks, you are still naked," Steph said. We held our breasts in our hands as we hauled ass down Harlaxton Drive during the early spring dusk with a feminist "YAWP!" Our only concession to cloth was the use of shoes.

The next week, we decided over fizzy pink Lambrusco to become roommates the following school year back in Indiana and continued our runs twice a week in shoes without socks. We enacted our feminist bravado with stinky shoes, a writing salon, and no shaving in the apartment we called "The hairy-leg café." We crayoned a poster for the wall so that those who visited understood. Steph even ran with me after I asked, "Can you keep your shoes outside? They make the apartment reek of rotten feet."

We ran and talked about the need for women to be able to run: Running away from the clichéd college area flasher who hid in bushes outside of our apartment building, waiting to open his coat and reveal his pathetic flesh when we walked outside. Running away from the nuns when we taunted the Catholic boys at the Catholic high school across the street from our apartment. "Hey! We want to fuck your sons!" we often yelled from the safety of our top-floor apartment as the boys scrambled into their parent's cars after school. Other times, we risked more and stood outside in our driveway hurling taunts like javelins, sure they wouldn't get our nuanced expressions. Once when the school was dark and shuttered for the day, we took a piss in the bushes outside of the school, running down Lincoln Avenue in a sober victory lap. Running *because* we were female.

Hung-Over Run
shuffle your miles
remember this knot of right
through the morning sludge

Spring/Fall 1991, Backpacking via Eurorail, Age 19, the Continent

In May, I get over Tim and the idea of romantic love as I backpack around Europe, mostly solo, after the semester in England. My friend Amy spent all of her money, and bailed out on being my travel buddy. I call my mother from the Harlaxton Manor phone booth with the last bit of money left on my phone card to tell her my plan and when to pick me up in Atlanta. "I just bought my return plane ticket. I will call you when I get in. Next week I'm going to the Continent!" I'm so excited that I am unprepared for her response. I thought she would applaud my adventure and my feminist bravado.

"Where are you going? Who is going with you?" She asked. I hear the money clicking away. *You have 5 minutes left for this call.*

"I'm going by myself. Amy spent all of her money and is going back home. But I'm still going. I'll meet Curtis and Printha in the Netherlands, though. We'll travel together for a bit," I said. I'm being defensive, and I know my friends can hear her concern as they stand outside the glass door in line waiting their turn to call home. I'm embarrassed.

You have 3 minutes left for this call.

"What is your itinerary?!" I knew she was angry when she commanded me, rather than asked me. My mother was usually subtle and gentle in her approach to mothering. I only have two destinations, as I intend to just look in my newsprint Eurorail schedule to pick and decide as I feel.

You have 2 minutes left for this call.

I quickly make up a list of places to tell her. "I'm going to Düsseldorf first to meet Gerhard's parents. Then, I will meet Printha and Curtis in Den Haag. They may be able to go with me to Rome, Berlin, Paris …"

"Call me." She said. I know that I won't as I hear the 1-minute warning telling me there is insufficient time remaining for the call. I don't talk to her until

16 Woman, Running

3 weeks later at a pay phone on the stone beach in Nice. I bought a phone card with just enough money to hang up on her again. I stare into the shockingly blue water and pretend there is static on the line when I don't want to hear the panic in her voice.

July 3, 2014: Bad Run
clutch your side stitches
summer ripened port-a-pot
eat a mouth of gnats

Summer 1992, DeKalb County Gun Range, Age 20, Lithonia, GA

My father takes me to the local gun range to shoot the gun I don't want, the small pistol that he bought me to take back to my college apartment in Evansville for my safety. *To protect yourself,* he tells me.

I tell him, "Dad, I can't take a gun to school. You know my roommate Athena would get mad at someone and shoot them." What I don't tell him is that my friends in ISC—aka the International Students' Club aka the International Drinking, Smoking, and Sex Club—who live in the dorms, hang out at our apartment to smoke pot because they are afraid of getting busted in official university housing like the soccer players did. I'm pissed that they do not think to put a towel under the crack in the door to keep the odors inside the apartment, so that many times I can smell who is at my place before I even make it up the stairs. Yes, a gun in my apartment would be the antithesis of safety.

When we ask for a practice target the clerk says, "You and your son can have the first lane." I look down at my summer uniform of jean cut-offs and striped T-shirt, shrug, and put on the shooting earmuffs that deaden outside sound. I can still hear Dad's gun safety lessons on repeat from childhood. "Never point the gun at anyone even if the safety is on. Always point the gun away from you when loading and unloading."

I hate everything about this; the feel of the gun in my hands, the weight of the bullets I load into the cylinder, cocking the hammer, the kickback when a bullet drives down the barrel and out of the muzzle. I can't use the sight to aim for the bulls-eye. Most of my shots don't even make holes on the outer rings of the paper target.

We don't return to the range.

When I return to Evansville sans gun for my final year of college, I spend no time with Tim or any of his fraternity friends, because they seem provincial compared with my international friends. I compete in the intramural run for ISC because most of the club members smoke, even in the dugouts during intramural softball games. And they never run. The ISC president signs me up as the best bet for the run around the dorms because I, at least, run sometimes. Plus, I had placed in the run for our team my first year of college. But my senior year pack-a-day smoking makes me wheeze around the dorms this second time, lapped by the sorority women who do not hang out with the smokers. I only finish that race because of my stubborn need to cross the line.

May 15, 2014: Last Run
like this is the last
like this is your dead last run
like this cigarette

Fall 1992, Lincoln Avenue Apartments aka ISC Village, Age 21, University of Evansville

I am studying in the third floor apartment I share with my best friend, Athena, when the Malaysian sisters who live downstairs open the door and shout, "You gotta come. Now. Athena and Cinar are fighting outside."

I drop my book, run down two flights of concrete stairs to the parking lot of the 15-unit apartment complex cursing because I have an exam that I'm making flashcards for, and this intervention will take time. I seem to be the only one in ISC who studies. (When I graduate with honors, they are all shocked because they never noticed me sneaking away from parties or getting up early to study.)

Here is the pattern: Athena drinks. Someone says something offensive. Athena bites the offender. Sometimes they bleed. Always there is a mark. The bitten and/or their posse find me and tattle. I am supposed to do something. Defend and explain.

So, I run to the scene in my role as handler. I see a circle of bodies, but I can't see Athena or Cinar. I just hear the shouting, "Athena punched Cinar! Athena punched Cinar!"

I can't see the punch through the lake of bodies lapping around my friend, but I hear Athena yelling at Cinar about our friend, Kris. The details of why she is pissed at Cinar are not clear, whether it was rape or regret. I'm annoyed because this is Kris's call, but we have taken the oath of women's friendship. We help one another make sense of the chaos. We have each other's back.

I remember this punch when Athena and I sit at our kitchen table with the massive and hideous ashtray I made in ceramic class with the mottled turquoise glaze I couldn't get right between us. We are smoking and talking. I confess after too many beers, "My dad bought me a gun, but I told him I couldn't have it here. You would get drunk and shoot someone." I'm not sure why I tell her this now, and why I'm nervous feeling like the *Rational White Girl* pointing out erratic behavior.

My aversion to guns is reinforced when Athena smirks at me, slugs down the rest of the Sterling Light, and says, "You're right."

First Mile
legs leaded down tight
hips fisted like a wish-bone
sleepy muscles bite

Spring 1993, ISC Spring Formal, Citizens National Bank Building, Evansville, IN

Athena, Kris, and I agree to clean the Homina Huminas'* apartment on the bottom floor of our apartment complex for a case of beer. Most likely we negotiated for Miller as that was the *fancy* beer. We did have rules, though. No picking up underwear. No cleaning toilets. We would dump ashtrays, pick up plates and cups with unidentifiable dried foodstuff, maybe do a few dishes, and empty the trashcans. Totally worth it, especially when I *borrowed* the pellet gun I found under a newspaper while we were there.

The week before, the Malaysian sisters called the apartment while I was studying, "Do you have any flour?" I told them, *of course. I'll be right down.*

I thought they were cooking, until I walked outside with the flour. The Homina Huminas were shooting their new pellet gun. At people. Correction. They were shooting their new pellet gun at the female residents of the ISC Village. Nuyen grazed my ankle when I leapt to the side of the air-conditioning unit to avoid his shot. It was a real pellet.

That fucker! He was likely mad that I called the police last month when he was outside in the parking lot yelling obscenities and misogynistic trash at his American girlfriend.

Athena and I help the Malaysian sisters make flour bombs in plastic bags to throw at Nuyen. We take a cup of flour, dump it in a bag with a cup of water, and then heave the bomb over the railing with righteous anger. When we hit Nuyen's maroon Porsche 911 with the flour and water mixture, he stops shooting and shouts obscenities at us as he races to the carwash to save his paint job.

The next month, Athena and I share the elevator with Cinar and Nuyen on the way up to the top floor of Citizen's Bank for the ISC Spring Formal. We borrowed dresses from friends' closets because this was the last formal we would attend before graduation, and we needed to look good. I wear black, and Athena wears green; we have fun plans for the night.

The weight of all our interactions the past year make it a long ride. I can tell they want to be in any elevator but this one. I wait for Cinar and Nuyen to look me in the eye. When they finally focus on me, I nod at the black satin drawstring handbag wrapped around my wrist, the one my grandmother had made for my high school Senior Prom. I reach my right hand into the bag, and slide just the butt of the gun out. They stiffen, and I enjoy seeing their pupils dilate. Their fear is delicious. They think I have a pistol, and they have no idea I stole the pellet gun from their apartment. The rest of the night, I relish the distance they keep from Athena and me, and the frightened looks they shoot me. I am powerful.

*The title Homina Humina arose when my friend Elizabeth made a comment about the male Turkish students' tendency to talk negatively about women in Turkish when we were around. I now recognize we used this xenophobic label as feminist avenger resistance strategy.

Feminist Run
run as strategy
physical mouthed resistance
strong-sweat out the shoulds

Summer 1993, Pizza Hut, Age 21, Gwinnett, GA

I discover that David drives around in his Toyota Corolla with a gun under the passenger seat on the way to dinner at my parent's house. Just in case. Because he is Iranian and misunderstood. Because we live in a racist, xenophobic culture. Because he is a cook with a temper and paints abstract acrylic landscapes to hang in his gray apartment.

David, my summer fling before my first year in graduate school, calls me at work at Pizza Hut from a payphone across the street: he was fired for his anger issues and is no longer allowed on the premises. I don't want to talk with him, because he is a terrible lay, and misogynistic, and I'm bored.

"Tell him I am not here," I say.

"Tell her to answer the phone. I see her inside," he demands. David who drove around with a gun under the passenger seat in his car. Just in case. He is across the street at the gas station payphone, watching me working inside. Watching me talking to customers and bringing them sweet tea and supreme deep-dish pizza. Watching me refuse to answer his call.

I have to figure out how to break it off with him, the man who drives around with a gun. Part of being safer is recognizing when things could turn ugly and learning from experience so that the body remembers and can react from that embodied place, understanding the intersection of white privilege and female disadvantage. How do I swallow my feminist convictions and play the jilted girlfriend role to get him to back off and move on to the next girl?

Injured Run
your legs will fall off
stop breath lungs gulp feet cement
bone crush joint pop ouch

1994, 5K Spring Sprint, Age 23, State College, PA

In my master's program at Penn State, I find running again. I become a consistent runner because of my roommate, Cheri. She is a runner. Her father is a runner. I decide that I want to be a runner, too. We lace up our shoes and run 2–3 miles

20 Woman, Running

together a few times a week. I get a subscription to *Runner's World*, and read the articles to learn what being a runner means. I go to the fancy running store on South Allen where the clerk watches me run down the sidewalk and back, diagnoses my over-pronation, and fits me for a pair of Etonics, the only brand of shoes I will run in during my twenties until they are discontinued.

Cheri and I run a 5K Spring Sprint sponsored by the hospital. Standing in a pack of runners before the race not knowing what to expect, not feeling like a real runner, I learn the nervous energy is best channeled into running in place, picking up each leg and pretending to do a high-step march like in band practice. Shaking my arms like a second grader playing the Hokey Pokey also helps. As we run past the test swine and test bovine pastures in Happy Valley on this impossibly bucolic dewy morning, I decide to quit smoking, even recreationally, because I like the running highs better than the nicotine buzz.

2014: Spring Run
reveal your new skin
too soon iced over feet, thighs
the snow will return

1997, 5 Miles, Penn State Golf Course, Age 25, University Park, PA

I mostly run solo the first 2 years of my doctoral program. Sometimes my roommate Joyce laces up with me, but she dislikes running. She also dislikes my refusal to quit Penn State and move in with the Russian scientist I date and fight with long distance. My feminism and focus on my career irritate her. "Do you want to end up alone? You could find a job in Princeton. You should move there," she argues with me weekly. Joyce is 8 years older than me and wants to be married.

"I don't want to get married. I don't want to have kids. Besides, he is the one who decided to move. Why would I quit my program?" I said ad nauseam.

"When I was your age, I thought that, too. At 33, I don't have anymore time. You don't want to regret losing your relationship," Joyce said.

What I regret is having to wear pants in the apartment because of her parade of potential husbands. This restriction is like when my previous roommate's boyfriend, Skip, moved in without asking; men before roommates was annoying. I did miss Skip's sharp knives, though, when he and Coby moved out to get married after their graduation. I remembered those knives when I buy a set of Henckels as a wedding present for myself 8 years later.

I decide to enter another 5K race by myself, because that is what runners do. I finish that 5K in a fast-for-me 24 minutes the weekend after I turn in my comprehensive exams that make me ABD—all but dissertation. All I think about during the race is running, not the week of writing and eating cereal for every meal and editing and searching through stacks of articles and books for the right quote and my aching wrists and back that only feel better with short run breaks. I didn't have to articulate the connections between theory and communication. I just ran like an animal.

So what if I lock my keys in my car before the race? I use a coat hanger I find in the YMCA closet and fish it through the window gasket to pop up the lock with my wrist and flushed limbs. In 2 minutes. Running gives me superpowers.

I run maybe 10 miles a week in 2–3 mile increments, until the afternoon I happen to pass Lynn, a fellow doctoral student in Education and Women's Studies, and her friend Naomi on the Penn State golf course. I had been jogging in the opposite direction, but I turn around and run with them. We talk about graduate school and writing and how we should start weekly women's dinners. We can host them at her place; her roommate who is in Wildlife Science and her friend in Education will join us, as will my friend in Communication Arts and Sciences. And we run. We run 5 miles—the most I have ever run at once. And it does not *feel* like 5 miles.

Lynn, Naomi, and I become running buddies. I discover a love of running long distances; I may run slowly, but I can keep going, especially when I have other women to run with. The battle is between my stubborn desire and the realities of the physical strains of running. My superpowers are confirmed when Lynn and I begin training for our first half marathon. Another bucolic day in Happy Valley, we are running along a trail in a new housing development when we pass Joe Paterno, beloved Penn State football coach, speed walking alone in his training flats and wearing those distinct chunky dark-framed glasses. As we run past him (and not just the cardboard life-sized cut-outs that adorn most stores in State College), he says, "What a fine pair of athletes." Joe Paterno (!) called us athletes, not something that I have ever considered my nerdy self to be. Lynn and I wait an appropriate amount of time for him to pass, and then high five. I consider that maybe he is right given I'm running 20 miles a week.

Shortly after my increase in mileage, however, my feet start to hurt. They crack when I wiggle my toes and ache most of the day. When I wake up in the morning, I have to hold onto the sides of my dresser and the wall to limp to the bathroom. I have always had bad feet. When I was young, my feet turned in, forcing me to wear hard-soled Mary Jane's that only came in red or blue to correct the deformity. This, of course, meant that boys teased me for not wearing the popular K-Swiss shoes. I stopped the bullying with hard kicks to their shins under the lunchroom table; there was no way they would tell on me with tears of pain in their eyes. A girl made them cry! I finally go to the health clinic and a podiatrist who the nurse practitioner has recommended.

I call my mom to whine about my foot pain and the second-degree infection, maybe Lyme disease, on my arm from an insect bite. My parents do not answer the phone, so I sob a litany of maladies into their answering machine when it clicks on: I will never run again. I have permanent damage. I will never finish my dissertation.

When my mother calls me back, I tell her that not only have I visited the health clinic for foot pain, I have managed to get an insect bite that resulted in a second-degree

22 Woman, Running

infection, and I have not written anything on my dissertation in months, the physical manifestation of school stress all over my running form. I will never run again.

"Did a doctor tell you that?" Mom asks the practical question. I never cried in front of my parents, because we were that white and stoic. Only passive aggressiveness was acceptable, so I knew my message was alarming.

"No," I say My sorry-for-my-selfness lasts 2 brief days. My parents help me pay for a podiatrist visit; I am diagnosed with plantar fasciitis, fallen arches, and osteoarthritis. Mom helps me get good shoes, orthotics, and plan new running routes.

July 13, 2014: Long Run
miles of miles of miles
sweat dried crystal salt thigh rub
angry red bra chafe

September 28, 1997, Philadelphia Distance Run, Age 25, Philly

When I try to remember my first long distance race, I find pictures of Lynn and I at the start and at the finish of the Philadelphia Distance Run, in a photo album from graduate school. I do not see any race numbers or medals—these were the days when only those who placed in a race received a medal—just flushed faces and the ugly red T-shirts with turquoise letters that show me we finished the race. This race was my first half marathon, my first long race after I discovered I could run and run. A stubborn girl became a stubborn runner. And though I'm a qualitative researcher, I do practical word problems, math that makes sense, in my head on runs: 20 minutes/4.5 songs= 2 laps around the golf course. During my late 20s, running and training for races keeps me goal-oriented, stubborn, and risk-taking in other areas—finishing graduate school and a dissertation, quitting a tired and too-long relationship, starting and then resigning from a tenure-track job, dating a woman for the first time, rediscovering poetry. Others see me as a runner.

Running reminds me to take care of self, to know when to ignore fear, when to run toward it, and when to run away.

Farm Run
move like a rabbit
through the origami hills
feet crunch on silage

Summer, 1998, Age 25, N Allen Street, State College, PA

When I return from Christmas Break, Joyce has moved out and into her boyfriend's place. She calls to reassure me that she will pay rent until our lease is up in

July. But, she is looking for a sub-letter. "If you find anyone, let me know." I don't search for a new roommate, because I enjoy walking around the apartment without pants. Though my joy is tempered by gendered guilt when I receive her half of the rent in the mail. She was a good roommate. We are not best friends, but we did respect one another. I washed my toothpaste out of the sink, and she split the grocery shopping with me. She threw me a 25th birthday party and made a cherry upside down cake from her mother's recipe. She took me to her parent's place in Bellefonte for Sunday dinner a few times.

This guilt makes me ignore my intuition when Joyce calls in March. "Rain is finishing up some undergraduate classes and needs a place to stay." *Rain? That cannot be her real name.* The more Joyce talks, the more wary I feel. Not paying rent would help Joyce finish her undergraduate degree and start graduate school in neuroscience sooner. I know the moment I agree to let Rain sublet my apartment it is a mistake.

Rain's boyfriend, Skank-Boy, moves in while I am away during spring break. The pattern of boyfriends before roommates seems to be another curse. I can't remember his real name now, because I renamed him after his repulsive behavior. I'm pissed because he's not paying rent, and he and Rain smoke so much pot that I have a permanent contact high. I can't invite friends over, because they are likely to sit on a pinch-hitter or have to sweep the stems and buds off of the coffee table to put down their teacup. I'm not writing my dissertation, and it is convenient to blame living with a clinically depressed woman and her clichéd druggie boyfriend. Skank-Boy talks slowly and refers to everyone as "Duuuuuude." I wonder if I can be jailed for knowing about their drug use. I only admit to myself they were shooting heroin in Rain's room when I call the landlord after Rain moves out to ask for the locks to be changed.

One evening when I return from a run, I see a cloud of weed smoke in the stairwell as I choke my way into the apartment. The door is wide open, all of the lights are on, and my best frying pan is burning on the stove. Rain and Skank-Boy reheat everything on the stove, because they believe the microwave is toxic. It seems that they forgot about the stove being on high with my pan as they took their high selves outside.

"Oh soooorrry. Duuude, we'll buy you a new pan," Skank-Boy says when I confront them later. They missed the point, and the fact that the apartment could have caught on fire with my half-written dissertation inside. I tell them that I put out a fire on this stove 3 years earlier when Cheri failed to notice a plastic container stuck to the bottom of a saucepan. I miss the opportunity to ask them to move out when I offer them some of the pasta I was cooking for my own dinner.

I feel more frazzled and ineffectual when I have to stop running for a month to let my plantar fasciitis heal. Rain and I go to the rec center and swim together. I try and talk to her about her depression, but this is the summer of my meeting no goal. I'm a failure.

24   Woman, Running

A month after they move out, the state police call looking for Rain. My new roommate, Andrea, asks me what is wrong, because our neighbor saw the police car pull into our driveway with the flashing lights on while I was at the library not writing. *I WILL NEVER FINISH MY DISSERTATION.* It is difficult to write about sexuality and sexual talk, but I'm not sure that is the reason I'm blocked. My interview research on Latinas' sexual talk in the context of their relationships with romantic partners, friends, and family is more than depressing. Listening to women talk about the messiness of negotiating their sexual desire and identities in a misogynistic culture is what is so awful. I'm holding their brave stories inside my feminist core as I witness the crappy relationship that is suffocating Rain.

"No, officer, I have not talked to Rain in a month. I have no idea where she is," I said. At least I could still lie. Rain had called the week before wanting to move back in. I told her, "I'm sorry. No more than two people can be on the lease, and my new roommate is staying." *Andrea begged me not to let you move in. And she does not have a loser boyfriend or do drugs in the apartment, so I won't get kicked out of my doctoral program.*

The next week the phone rings with another unwelcome caller. "Heeey. Remember meeee?" Skank-Boy remembered the phone number. I had the locks changed, but I kept my phone number. "Would you like to maybe like go running sometime?" *Did this dude who cannot run because he smokes so much pot that he wheezes walking up stairs really just ask me out, his girlfriend's former roommate?* I need to take a shower to wash off this feeling.

Cat Calls
don't call us a girl
don't call us a girl jogger
fierce women running

May 19, 2000, Girl Bar, Immersion Ethnography Class, Age 28, West Hollywood, CA

It takes me two warm-up laps around the block and a quick coke in a dirty diner to work up the courage to stumble into the club alone. I took a morning run through the Hollywood walk of fame, stopping to look at some of the bronze plates embedded in the sidewalk, as a stress-reliever, but my flight impulse makes my stomach sour now; waiting for my ID to pass inspection in front of the club feels like a lesbian ID card check. *Why did I decide to study lesbian space during these 2 weeks in West Hollywood?* I joked earlier with Jo, a volunteer at the Mazer Lesbian Collection, that I needed to show her my card to be a legitimate visitor. Jo informed me that the card joke circulates, but she never asked me, "Are you a lesbian?" This question I have been asked both explicitly and implicitly through the look or flirt test too many times to remember. She showed me the personal lesbian history tucked away from the rooms that hold more public newspaper clippings and novels without comment or question. Only the real lesbians get to see these letters and love notes.

After I pay the dance club's entrance fee and get the stamp on the back of my hand, I make my way through the steel-gray carpeted lounge area to the bar. I'm meeting Mila here at the Friday night Girl Bar, the revolving lesbian dance space at the Factory Club in WeHo. She asked me if I would meet her for a drink after I nosed into her conversation with a clerk about girl bars yesterday at the LGBT bookstore. Because I was 4 days into a 2-week ethnographic study of lesbian space, I knew all of the gay hangouts. I expect Mila to appear around 11, so I have time to write notes about the scene. A metal cage in the center of the room divides it into three sections: comfy chairs and couches, dancing cage surrounded by stools, and the bar. This place reeks of eroticism—the cool décor with dove grey walls, pearl grey pleather overstuffed couch, low barrel tables, steel support beams, and ceiling tiles, tiny lights on curvy metal rods, heavy grey velour curtain dividing the lounge and dance spaces—as do all of the women in it. I step around the steel-rod bars to the shiny counter, tugging my too-short lace top over my black rayon skirt with the too-high side slits. I think the lacy t-shirt looks like a box of crayons vomited on it, but when we were costuming for the night at the Motel 6 my classmates convinced me that it radiated hot club vibes. I had to take a shopping trip for field clothes, since my suitcase was full of my usual uniform of cut-offs and striped boy t-shirts. *How could I have predicted what I would need for this immersion ethnography class back in North Carolina?* I hadn't considered what I would study or known what participant observation meant here in the land of lipstick lesbians.

When the bartender asks what I want I order a coke, not a beer, because I'm working. Alcohol would soothe my anxiety, but I need to be sober Sandra, the earnest, fresh-out-of-graduate-school communication researcher. "Would you like a cherry? A lime? A lemon?" The bartender wears a black chiffon turtleneck with a black lace bra underneath and black jeans. As she pours my coke into a glass, I discover all of the bartenders are women dressed in the same uniform.

"No, thanks," I say to the hot bartender. All of the bartenders are hot, of course.

I want to get away from this interaction, away from hot bartender #1 and her straight, raven-black hair that gleams under the TVs suspended above the bar, which are playing geometric patterns, swirls, lines, and squares of color in slow motion. I feel as if I will be laughed at, turned away, and punched in the arm by the bartender and everyone else in the club because I'm not gay or pretty enough. Going to a bar alone, especially a woman-centered one, frightens me more than my first day in graduate school, more than the first class I taught as an assistant professor, more than telling my friends that I hooked up with my new female colleague. Maybe I'm not ready for all of the raw eroticism. I can't walk down the streets in Greenville, North Carolina, and hold my girlfriend's hand. How can I act suave and sophisticated when it requires pretending I see this many attractive women who are openly into other women every day?

26 Woman, Running

There is a female couple making out against the wall to the right of the bar. My new outfit purchased for tonight does not work. I do not blend into the lipstick land of WeHo. And it isn't just because I wore the Dansko clogs, which are comfortable and decidedly unsexy. I laugh about feminist scholar Suzanne Pharr's notion that "lesbians look like all women and all women look like lesbians." Most people in college assumed I was gay, especially the men I refused to suck face with at parties while I danced with my female friends during slow dances. "Are you a lesbian? I hear you date (read: sleep with) your roommates." The price was lesbian baiting, being called lesbian or dyke or feminazi when I didn't act straight enough, didn't flirt appropriately, and when my behavior threatened male privilege. The goal is to push women back into subordinate positions. Using lesbian to label a woman acts as a control for all women in that no defense is possible. The refusal of male authority and approval means being perceived as lesbian, to lose heterosexual privilege because of unacceptable behavior. Because sexual identity is often (in)visible. I was not dating women then, but my crimes were independence and mouthiness, refusing to be a straightforward subordinate flirt. I got dates with men by making fun of their clothes (the Russian scientist), their religion (the Jesuit School student), their problems with feminists (the French work-abroad student in London), and traditional gender roles (the older artist). Insults meant I left with phone numbers or their hands down my pants. It is difficult to prove sexual credentials. In the food line at a house party, I told the scientist I knew he wasn't from the US because of the stitching on his loafers. We argued about the relevance of empiricism for the study of relationships once we got past the "what do you study" small talk. "The boiling point of water does not matter!" I remember yelling, so that other conversations around us stopped, pissed that he was insulting my interpretive research approach. I got his phone number on a paper napkin. I left our 5-year relationship to date Kathryn. In the closing relationship arguments, he accused me of sleeping with my best female friend in graduate school, yet his last words were, "You're not a real lesbian."

A Runner-Runner
a real runner runs
no jogging or walking or
skirts and socks that match

November 18, 2001, Philadelphia Marathon, Age 30, 358/507 Females Age 30–39

Turning 29 bothered me, and the bothering over it bothered me as an out feminist, so I trained for and ran the 2001 Philadelphia marathon as a 30th birthday present to myself. That summer was emotionally and socially like a series of violent thunderstorms. I broke up with Kathryn, who was in my small professional circle; ran my first 10K with the best bib number ever—69—safety-pinned high on my chest; and let myself flirt with the bad boy in poetry class: 15 years older/ divorced with 15-year-old daughter= bad feminist. When Paul moves in with me the following year, everyone is disappointed. I don't find out until I break it off with him 2 years later lying in my bathtub in Syracuse while he sits in our duplex in State College. I am thinking about how pleasant it was to go apple picking with Josh, how much better I like myself when he's around.

"Is this like a divorce?" Paul asks.

"Yes," I say. I am careful not to splash the water. I think we both knew the relationship had been dying the slow death of long distance when I moved to Syracuse without him, but it felt disrespectful to break it off in the bathtub. I find out that Paul takes up running after our split.

That summer training was my only steady. I ran the Philly Distance Run again as a *training run*, but the run was tough for me that day. I felt like an uncoordinated elephant calf as I plodded through the course, which made me ask a series of questions as I continued to train the remaining 2 months before the marathon: *Was I running enough? Could I finish the race in 4½ hours like I wanted? Was I really a runner?*

During the marathon, I muttered a mantra to keep going—"this is what 30 feels like." What 30 looked like was 4 hours: 37 minutes: 6 seconds (4:37:06). I ran 7 minutes slower than I wanted, but the point for me: I ran. I finished. And then I quit running. It was not intentional; the rest of my thirties were spent sporadically not running. I would take a run, and then months would pass. I would run a mile or two as a warm-up for karate. I blamed my identity as runner to non-runner on the torn ACL from a spar in karate class and the surgery a year later, taking up snowshoeing and hiking, the hills in Syracuse, a semester teaching in Madrid, the pregnancy and birth that wrecked my pelvic floor. Of course, some running books perpetuate the idea of woman runner as fragile, the feminine body as caregiver, mother, other. What these books fail to mention is that running serves women as a reminder to thank the body, to discard damaging notions of femininity, and to rework the self from the inside mile after mile. Feminist identity is good for relationships.

Run Body
thunder with your thighs
running is like being born
saying yes to cheese

July 29, 2005, Wedding Planning, Age 33, Syracuse, NY

"What are you doing on Monday?" I ask my friend Di. I then call Bern. "Josh and I are getting married and need a witness. Meet us at the courthouse at 12:30?" It is the last Friday in July. I should be leaving for Madrid to teach for a semester, but instead, Josh and I plan our wedding over the next 3 days. We were hoping for a ceremony in Spain, but the wait for my work visa crushed that idea.

We will get married at the courthouse in Syracuse, New York on August 1 after no engagement. The date is not significant; when we call officials listed in the phone directory, Judge Aloi is the one who can marry us on his lunch hour.

On Wednesday, we had gone to the courthouse to get a license. In New York State, you can get married 48 hours after filing the paperwork. No blood test needed. When the clerk asked what names we wanted to list on the form, I turned to look at Josh. "Want to take my name and be Faulkner?" I asked.

"No," he said. "Want to be Atkinson?"

"No, I like my name," I said. I've had this name for 33 years.

We call our parents to tell them we're getting married on Monday. They're not surprised. Josh had dropped the lease on his dusty apartment to move into my house in May. He was my work friend for Thanksgiving at my parent's place, and I was the only woman he had ever introduced to his family's dairy farm in Missouri. They heard us puzzle about getting married in Spain because same-sex marriage is legal and that fits our politics, identities, and social circle. But the me getting to Spain part was being postponed, and making a legal bond there was expensive.

We would get married in New York. I would go to Spain without him. He would stay in Syracuse to teach and care for our pet rat and house.

My mother asks me, "What are you wearing?"

"Uh, I don't know. I guess something in my closet—hadn't thought about it," I say. As someone who never intended to get married, and definitely not to a man, I also never envisioned a wedding. No dream day woven into a romance narrative of womanhood. Marriage meant having to iron some man's shirt.

On Saturday, we wake up and consider my mom's question. We buy new clothes from the trendy bohemian thrift store around the corner from our house.

On Sunday, I decide I want some flowers. When we walk around the block to the Westcott Florist, we see the sign: CLOSED. Okay, I will carry a grocery store bunch of red confetti roses we buy at Wegmans. I find some golden ribbon in my craft drawer to string together the stems and place the bouquet in a mason jar of water. Di and Red will pick us up and drive us to the courthouse (only a year later would I realize it was the criminal courthouse, as we drove by on an errand).

Sunday Run
chained shut factories
resting quiet place to pee
outdoor service time

January 2011, Mother Running, Age 39, Toledo, OH

My virtual 10-year running hiatus was over in December 2010 when Marne, a serial marathoner student-turned-friend, made me a training plan; I went from not running to doing a 5-mile leg of the Glass City marathon as part of the BGSU Women's Graduate Caucus Relay team in 5 months without too much physical pain. I took 3 weeks off in January when my out-of-alignment hip strained my IT band, which messed up my gait and inflamed my tendonitis and so on. Most of my training pain was psychic and identity related, as I contemplated turning 40. It seems that the last year in a decade finds me wrapped up in my gendered training about the worth of older women. I alternate between being excited and horrified at the idea of cronedom.

The first wobbly walk-run lap around the City Park I feel damp after a few steps. I look down at my crotch and notice a huge wet spot. *Did I pee my pants?* I waddle over to the restroom and confirm my shock. *Thank you* childbirth. *Will this happen every time I try and run?* At least I am wearing black running tights. That day, I turn around and slump home, disgusted and ready to never run again. The other days, I skip my morning cup of coffee before a run to encourage my leaky bladder to take a hiatus. This works only 2 days out of 4, so I work on being less concerned about containing my leaky middle-aged body.

Though my maladies have multiplied during my running break—tendonitis in the ankle and knee, weak pelvic floor, IT band strain, inflamed plantar fasciitis—I continue to walk-run until I can run 2 miles without walking. I want the feel of being a runner again; the aches and pains written in my spidery veins and knotted tendons. All runners get injured. All runners will be injured at some point in their running. For women, the equation of hormones + locomotion + impact mean our bones and tissues absorb 80% of impact as feet collide with ground 5,000 times during an hour run. But what mattered most is that I regained the feeling of running, what it does to stabilize and rework a middle-aged female body as strong and capable, like sweaty magic.

Running @ Home

I run through my woods
like a dog on her patrol
I urinate here

Spring 2012, Nurse Sandra, Age 40, Lithonia, GA

I enjoyed the training runs best the week in April I took off from being a mother, a professor, and a partner to spend with my dad, who had 4 weeks earlier had his right leg amputated at the age of 71. Running on the trail through the Georgia pines where I grew up with my dad made me feel grateful that I still had my strong legs that look just like my father's leg(s). I got to mourn his leg as I pushed up a hill gulping the scent of honeysuckle, wound around a bend dodging pine cones, knowing it was easier than mourning a dead father. It also made me acknowledge being a middle-aged mother. The most dangerous thing I do now is run alone over slippery composite wood planks while not thinking of the scene as bucolic when I enter the mist at the top of the hill by the babbling creek and mossy rocks; I think of how *this* is the serial killer view and I am woman bait. And I seek out this view.

Forest Run
only your head reaches
the pine tree knuckles, you drag
the ground with straw claws

7/26/12, Age 40, AAUP Summer Institute, Chicago

"Can I tell you something without offending you—no wait, I shouldn't," Rob asks
and unasks in one breath, leaning close to my right ear to be heard. We are drinking
in what I think is an Irish bar close to the Palmer House Hilton in Chicago. Earlier,
it was shuttered when I passed by on a morning run. Now, we found no empty
seats or tables, so we are standing by the dark wood door and clover-green banners
advertising beer with some of my colleagues from Bowling Green and members of
the American Association of University Professors (AAUP) national office.

"Go ahead. I won't be offended," I say, and take a gulp of my Manhattan to indicate my goodwill. When I walked into a conference workshop late a few hours earlier, I recognized Rob four rows back. It was as if we were still office mates at Penn
State, kvetching about finishing our dissertations. Though now we are both grievance
officers in our universities' faculty unions, growling about the state of the profession.

"My wife thought you were a lesbian," Rob says out of the corner of his
mouth, arching his eyebrow. We are just an afternoon into the reunion and have
not exchanged the 12-year timeline of life events, just the expected soccer picture
of his boy and the pre-school picture of my girl.

"I'm not straight!" I shout. My tone is inappropriate, but I'm frustrated. Rob's
statement about sexual proclivities makes my petulant, adolescent self puff out,
ready for a knife fight. I started dating Kathryn after Rob and I left Penn State
with completed dissertations, but then there was Paul, the 15-years-older-than-me
man I lived with for a year, followed by my husband. I can't unravel the pattern
when my sexual identity is a question, when it's important to tangle or unravel the
non-linear graph of who I sleep with and what and why I love. Why do identity
markers hinge on who is my partner at a particular moment rather than on me,
the bisexual person?

Women Running
safety dressed in form
Caution: RUNNERS on the road
dogs catcall danger

August, 2012, West Hall, Age 40, Bowling Green State University Campus

My 40-year-old self still wants to be asked to the gay prom. I have returned to my
field notes from the West Hollywood ethnography class in 2000, digging inside

my moldy office to find the manila folder with the descriptions of conversations, phone numbers, and playbills. I read in spiral bound notebooks about my annoyance when asked about being a lesbian. These questions about my status as lesbian were all asked in lesbian spaces when I was dating a woman. I smirk with a little thrill at the collection of numbers, though, and the memory of getting bitten twice by two different lesbians in two different bars. I remembered the awkwardness (and not the skill) in playing the novice researcher.

There are also the other moments in fieldwork when implicit questions about my identities necessitated the creation of an explicit script. Recruiting interviewees for the narrative study on LGBTQ Jewish American identity I completed with my postdoc mentor, Michael Hecht, required: "I'm gay. He's not. He's Jewish. I'm not." I identified with the bisexual lesbian; a label she chose to honor her past relationships with men and signal a shift in desire. Identity labels are political by necessity, even as experiences swim (drown?) in the Sea of Flux. Funny how now the people I meet usually perceive me as queer, how I have ended up mentoring women who love women: four graduate advisees, a roommate, a research assistant. I note the raised eyebrows or mouth twitch when my male partner and child become apparent. I guess it is like the ethnographer Tony Adams' claim that coming out is a process, a life-long series of decisions about how to perform as bisexual or semi-straight or queer-straight or anything but just another married chick with a kid and a dog. To me, I have made the queer choices that on the surface reflect traditional norms. Always, I must explain that I'm not straight, that my younger self never planned to be married or to parent, but my current relationship is based on equality and not misogyny. And that's hot. Yes, I have a male partner. Yes, I have a biological child with said male partner. But who and what I get to be with a lover is more important than genitals.

Vacation Run
salt water fogs rocks
birds dressed in yellow slickers
smells sound like your sweat

October 21, 2012, Detroit International Half Marathon, Age 40, US/Canada

I'm at the start line of the 2012 Detroit half marathon 4 days before my 41st birthday, as a present to myself. This is my turning 40 half marathon, my first since 2001, to celebrate the fact that I am a runner again. After completing the 5-mile segment of the Glass City marathon in April, I promised Marne I would run a half marathon with her. She makes me a half marathon training plan, and I figure out how to contend with my sieve-like bladder while running. A big concern in my running now is not ending up in the student newspaper blotter for public urination. My bladder is as leaky as gossip about professors and students I hear before

32 Woman, Running

class—who is in the newspaper blotter for public intoxication, whose class is boring and hard. I have to pee a mile into a run no matter how many times I use the porta potty before a race, even if I don't drink any coffee or water before I lace up. When Marne or other students run with me, I joke that I know all of the spots to pee in Bowling Green, and I stop to prove it. The low point would be the thawing winter day that I cannot hold my urine. I'm outside of my daughter's elementary school with my competing desires and go for the most immediate need. I finish my 5-mile run, amazed by how little attention people pay to middle-aged women peeing in bushes. I guess the fact that aging women become invisible may just be another superpower.

Marne and I finish the race together with a fastass-impossible-for-me-to-breathe last mile (8:43). We stopped and took pictures along the route, ran through the sunrise over the Ambassador Bridge into Canada, and agreed to run a race together once a year. I feel invincible having run to Canada and back. During the run, I pledge to myself that my daughter and I will run a race together. And we do.

We practice around our block; I stand 500 yards down the sidewalk, hold out my hand for the finish line, and my daughter sprints with her arms akimbo and shouts, "Wait for me!" The next spring, we run a quarter-mile race to raise money for the local domestic violence shelter. I have a picture of her holding her right hand up in victory, medal around her neck, and determination on that 3-year-old face.

We repeat this again the following spring. Run. Raise money. Sprint around the block. Take a picture of a victorious 4-year-old face in the Toledo Rocket Glass Bowl Arena.

After this race, my daughter comes home from pre-school with a family picture she drew. The narration: "This is Josh. This is Mommy. This is Mimi. Mommy and Mimi are running." And though I'm ambivalent about the identity "mother runner," my wish for Mimi is this: Run fast. Run strong. Run like a woman.

Charity Run
we run this for you
picture pinned to race day shirt
for you, you and you

Spring 2013, The Year of Five Half Marathons, Age 42, Midwest

I run five half marathons in 2013, not the one a year I had pledged in 2012. And the time I finish matters. I have turned into that runner, the one who cares about time. My declarations that running is one place in my life that I'm not competitive are lies. I do find running meditative, but I can't meditate my way into just enjoying being able to move. It's not shocking that I will get injured this year. Each race I shave off time from the previous one until I have to stop:

April 28	02:18:15	Toledo Owens Corning	10:33 pace
May 3	02:17:39	Cincinnati Flying Pig	10:30 pace
August 13	02:23:46	Bowling Green Boy Scout	10:58 pace
October 20	02:15:47	Detroit	10:21 pace
November 9	02:17:40	Church-Hills 44 1/2	10:30 pace

In the Boy Scout Half Marathon, I place fourth in my age group, this the race I wanted to quit. But runners don't quit. I'm surprised my time is not worse when I look up the results online at home after I stop having chills and nausea from dehydration. I take a cold bath, because I feel like I will pass out. Before the run, I did a half-mile kids' race with my daughter in the August humidity. The temperatures were in the 80s with 80% humidity, too miserable to run. Running in any temperature above 68 degrees Fahrenheit feels like running in a sauna with saran wrap and a weight belt on with a 100% incline. During the last 4 miles, I think about finding the sag wagon and admitting failure, but Marne talks me over the finish line. The last 2 miles, I have to stop every quarter of a mile to walk, which is not my usual mojo at the end of a race. I like to find a person slightly ahead of me and pull out my inner sprint to pass them to the end point. Instead, I shuffle along, every step stabbing the bursa in my right hip, my feet like baseball bats hitting the asphalt and sending waves of unpleasant knocks up my hamstrings and quads.

"Tell me about your new job," I beg. I want to vomit. I want to lie down on the grass in someone's yard under a sprinkler and never get up. I want to do anything but keep moving. Marne narrates the joy in her first tenure-track position, teaching women students, feeling like she is fulfilling a calling. *We are almost there.* I make it to the finish, because she stayed with me. *We are almost there.* She talked me into not quitting. *We are almost there.* I think later that running is feminist embodiment.

In November, my plantar fasciitis flared up because of the minimalist shoes I wore in the Church-Hills race, my fifth half marathon of the year. Also, the second time I ran two half marathons close together, which was most likely not the best decision. I can't jump off the runaway train of too much running. For weeks after, I hold on to furniture first thing after getting up from bed or a chair, using the leverage to make my legs stand on feet torn like so many little slips of paper. There is no way I can break a 2:15:00 half marathon time, a goal I whisper to myself during long runs. I didn't know then that my training as a relational scholar would be tested. I didn't know then that I would place fourth in my age group in the 10K at the Gay Games the following August. I didn't know then that my plantar fasciitis would worsen because of my running and hiking around Germany after the Gay Games on my sabbatical, forcing me to take the winter of 2015 off.

34 Woman, Running

Speed Work
tell your body no
not tired take over brain
take these legs and move

January 6–9, 2014, Knitting a Polar Vortex, Age 42, The Church-House, Bowling Green

"I'mcold-I'mcold-I'mcold!" Mimi says again and again and again. A piece of the polar vortex breaks off before the finish of the warming period, plunging into our corner of the Midwest because of the displaced polar jet stream.

"Put on a hoodie. It's winter and supposed to be cold," Josh and I both tell Mimi. We preach about the effects of global warming evidenced in this winter and ignore the tired and inane commentary about global warming as hoax and this icy never-ending cold as God's wrath for sinners like us. Josh, Mimi, Buddy, and I swaddle ourselves with a space heater, the furnace coughing to keep the temperature to 55 degrees Fahrenheit with the negative double digit wind-chills. The 35-foot ceiling to floor drop and the wood rafters that rise like a bat obstacle course in our church-house require layers of robes to fight off the evil wind.

Thus, winter break drags us along a 3-week loop of waiting. Waiting for the hard parts to be over, though I know impatience turns the days into excruciating hours that can only be endured: We wait for snow emergencies to be lifted, my term as grievance officer in the faculty union to be complete, and for Josh's primary care physician to fax the papers to the surgeon okaying him for the thyroidectomy.

We finally get it all scheduled a week before the new semester will begin. We will celebrate Christmas with our BG family at our church-house. During the surgery, Mimi will stay at their house, Buddy at the boarders. I will drive Josh to Toledo Hospital for the surgery and stay overnight with him.

The phone rings while Josh is taking a shower. This is the second day in a row that the shrill sound of disappointment interrupts our departure. We had been refreshing the Sherriff's office homepage, willing the Level-Three snow emergency to be lifted because only emergency and essential personnel are allowed on the roads then. This thyroidectomy is not emergency surgery, though for Josh and me it is, given what we have had to do to get to this point. He wants the cancer cut out, and all I want to do is take a run. But even my now normal three pairs of tights, two long-sleeved shirts, windbreaker, scarf, and wool hat can't block the butt-chilling wind. I know that I would feel better if I could go run. Neuroscience journals are full of research demonstrating that running helps with anxiety.

Level 3 means Josh and I are frozen in, alone, waiting to get to the hospital. We can't even retrieve the rest of the family because they are too far to walk to in the NW Ohio tundra. We have enacted the last meal supper two nights in a row, too. No eating or drinking after midnight means I cook something full of fat and hard to chew. We can't seem to consider the time alone as a couple getaway.

I have to answer the call from Dr. Wharry.

Woman, Running **35**

"Mrs. Atkinson?" Dr. Wharry asks.

"No," I say and let the confused silence hang on the icy line for 20 seconds. "It's Dr. Faulkner." I'm not making this easy for her, but I can't. Even though she made it easier for me when she told Josh and me that this thyroid cancer wouldn't kill him, something else would.

"I am Josh's spouse," I say finally.

I talk to Dr. Wharry and reschedule the surgery for the second time. We agree on Thursday and not the following day. Better to allow an extra day for driving bans to be lifted, especially because I'm sure she heard Josh swear as he paced on the altar. "*I'm never going to have this surgery done. I can drive on the roads, so why can't a surgeon living 2 miles from the hospital drive.*"

I decide to shampoo the carpets and make nachos for lunch with our leftover cheese. Josh stays huddled by the space heater, our frustration and disappointment a hot contrast to the chilled air; he reads and I knit what I will later call a polar vortex scarf. I understand how my mother and grandmother knit to ease time and the mind-numbing pain of waiting. *I understand now.*

The Sunday my father's aneurysms killed his right leg and he spent all night in surgery, states away from me, I made two pans of lasagna and stayed up all night editing a manuscript, thinking *I'm not ready for my dad to die.* When I spent almost 3 weeks in the Wood County Hospital because of unexplained bleeding before finally giving birth to Mimi, I polished my tenure portfolio, finished knitting Josh a forest green cabled sweater, and began knitting a red and orange merino wool button-hole shawl, thinking *I'm not ready to do this birth.*

And finally as I wait for Josh to get out of the recovery room after the thyroid-ectomy, I knit on a green wool hoodie sweater my mother started for me in college and talk to the others waiting for the buzz and blink of the pagers we were handed by the receptionist and think, *I wonder when we can say the word cancer-free.*

2014: Winter Run
put on the layers
sweat hot pink skin under tights
polar vortex breeze

Easter Weekend, 2014, Painting the Church-House Doors
Harlot Red, Age 42, BG

My neighbors walk-by with their kids and dogs properly leashed while I paint the doors of my church-house red. I'm the one who is painting because I've internalized the family oath of love as action and social support as doing. I get the irony that love is not verbal expression for this tenured communication scholar as I compare the old-barn-red color still clinging to the door frame trim with the pay-attention-to-me-now red that slops off my brush and spills down the outside of the gallon-sized can onto the front stoop. *Holy Wow,* this is brighter than I anticipated. I shouldn't have scoffed at the Ace Hardware employee who marveled at

36 Woman, Running

the paint's brightness when he thumbed a splotch on the top of the can and asked me, "What are you painting?"

"The front doors." I didn't say the *of course*. When you live in a former church, the *door as mouth* metaphor means there is no other color to decorate a door. Red doors signify welcome if you follow Feng Shui, though I prefer the Scottish idea of paid-off-mortgage-red and superstitious church folk's screw-off-evil-red. My lurid door color says here is a place you can take-off-your-shoes-red. Red means you manage the messy art of containment. Something I had been doing since my husband Josh's cancer diagnosis and thyroidectomy during the polar vortex that created an apropos colder than normal winter.

As my family ducks under the ladder on the front stoop on their way to the playground, I concentrate on not falling off and shiver in my camo cargo shorts, optimistic about the almost spring weather and the color I stroke on the fiberglass doors we had installed when it was too cold to paint. Mimi wants to help and yells at me from the sidewalk, "Mama, can I help paint?"

She is at the age when helping to do house tasks is fun, and I'm at the age when her help means I shout my frustrations in bad mommy voice. Josh is taking Mimi to the City Park so that I can paint in peace. I'm not wearing a mask or gloves despite the bold print on the can label—*WARNING: This product contains chemicals known to the State of California to cause cancer and birth defects and other reproductive harm*. I imagine my kid covered in the unctuous red chemicals and say, "Maybe later. I'm doing the hard part now." Josh pulls Mimi's hoodie sleeve to steer her away from the would-be disaster.

"Hey Josh, don't you think this color should be called Harlot Red?" I smirk to myself about my cleverness of naming and using such a brazen color to paint church-house doors over Easter Weekend, until I realize I'm going to be painting for days. I have violated HGTV's entire list of Dos: 1. DO expect to apply three coats when painting red; 2. DO buy only a quart first before you commit to the color; 3. DO wear appropriate clothing. I hadn't considered that the hue and intensity of the red color family would mean more work, even though my Uncle, the chemist, and my hair stylist, both tell me how difficult it is to get red to take. *I wonder if this means that red dye really does cause cancer? And since cancer visited our house once, does this mean it will return?*

I have to stand on the top of the stepladder to reach the apex of the doorframe, so when friendly neighbors gawk past and say hello, I focus on not shifting my weight too much. My legs ache from the effort and my weekly long run; I've increased my mileage too much in an exuberant show of how well I weathered the emotional labor of this winter of Josh's cancer and sub-zero temperatures. A hot afternoon bath followed by a nap would be preferable to the tedium of painting, but 4 months of primer-colored doors was enough. We didn't have the doors pre-painted because they already cost too much; non-standard size doors and the expectation of etched oval designs on fancy frosted-glass cut-outs are some of the price we pay for living in a former church.

I won the debate about which house we would buy when we moved to Bowling Green 7 years ago with this: *if we lived in a church, we would never have to go to church.* A redundant argument for an atheist, I know. The Archdiocese of Ohio owned it first, and constructed a small, dark, rectangular brick structure in the 1910s on a plot of land that would later be deemed the Boomtown Historic District. A neighbor bought the church from the local Nazarene congregation in 1973, painted the outside bricks white, and converted it into living space with skylights and a loft. Josh found the thought of living in a former church both repellent and fascinating. He researched the ownership history of the space, while I turned the distinctive marks of the former church turned family space into good omens—the scuffs, scrapes, and bolt marks from where the pews rested on the wood floors, the swirly beige and white stained glass windows that become orange flame shooters when the light beams in like a sun dial, the altar we re-carpeted with burgundy shag to soften the falls and dirt marks from play, and the feel of walking over well-worn grooves.

As I paint and paint, the most ubiquitous comment I hear is "Looks good!" But I've learned that in the Midwest, one does not directly voice their true opinion. So depending on who walks by, I respond one of three ways:

First response to those I don't recognize, "Thank you."

Second response to my immediate neighbors and those who I imagine will complain to the city about the color choice, "It's brighter than I thought, but I imagine it will look good once I get all of the paint on. At least it's warm enough to paint now." They pause their walks so that we can brag triumphant about how we survived the polar vortex that cycloned on from one level 3 snow emergency to the next. Think of the polar vortex as a large cyclone that circles our planet's poles with cold-core, low-pressure areas that rev up during the winter and back off in the summer, breaking up in the middle of March to the middle of May. We need a cataclysmic break-up with our vortex now. I paint our doors Dissolution Red in hopes that like a new hair color or tattoo after a breakup, this act of renewal signals moving on.

Third response to the few who I consider will appreciate my wit, "I think of this as Harlot Red."

Suburban Run
no way winding out
instead of church pray, run, here
lost in beige siding

August 9, 2014, Gay Games 9 (GG9), Opening Ceremony, Cleveland-Akron, Ohio

Being here is like getting to be an athlete *and* a nerd—I see older women and men and beautiful non-binary people all together and get to just be and don't have to shove myself into a gay or straight category, a toned 20-something young woman

38 Woman, Running

versus a slightly pudgy invisible middle-aged woman, a runner or a researcher, an athlete not-athlete binary—the either/or shot putted out of the Cleveland area. *And who doesn't love rainbow colors?* Lately, when I ask my daughter what her favorite color is, she says, "Rainbow!"

Andrea, my roommate at the end of my doctoral program, drove from State College to spend the weekend with me and attend the opening ceremony of GG9, since Marne can't. Andrea and I haven't been in consistent contact since graduate school, but she senses when to reach out with a call or email—like a *gay-frienddar*. She sent me a check-in email the week before. I bitched back that I'm stuck doing this research on women and running solo, because Marne would only arrive in time to run the 10K with me and then leave. I've spent the summer interviewing women across the United States about running, and now I get to run in the 10K and the half marathon at the Gay Games. I've been training and documenting that training all summer, and I want to kick ass as a researcher *and* a runner. I have a backpack full of consent forms and "**Women who Run**" recruitment flyers advertising, "All women identified runners: I'm interested in your running story!" I'm finally doing the research project I've thought about for years, and I don't give a shit if it fits into my academic program of research. I'm running toward the feminist ethnographer I want to be.

When Andrea arrives, I forget about my research role, and we focus on me getting to be a runner. I booked a hotel room in Akron for the week since all of the road races will take place there. Josh and Mimi will come and spend 2 days with me at the end of the games to cheer me on during the half marathon. At first, I had only signed up to run the half marathon, but I added the 10K to my agenda when I sent a picture for my athlete badge to confirm my registration (*oh how I agonized about what picture to put on the front of the badge*). I knew that being just a spectator wasn't going to be good enough. I would want to be running, and I found out that I could sign up for all of the road races with one registration fee. I'm doing participant-observation research on women's embodied experience, so I have to run, right? Running is research. Running as research. This bravado makes me pick up a race number for the 5K when I see the short line to claim a number as I get my 10K race bib. Since I'm doing research, my running times shouldn't matter. I dislike 5Ks, because I'm not a sprinter and don't feel anything like a runner until I'm 4 miles into a run. My favorite run is 10 miles, but there aren't many 10-mile races. A half marathon is 3 miles too long, and a 5K is 7 miles too short. Now, I will run three road races in a week—36 kilometers= 22.4 miles.

Andrea and I figure out how to take the shuttle bus from our hotel in Akron to the Athlete Village in Cleveland. On the bus, we meet a fellow runner, Chuck from San Francisco, and talk about women and running. I don't think about how sexist his comments are until later in the week, because I can only think about registration and my excitement at being here. Andrea and I take our picture together under the GG9 banner at registration where I picked up my athlete participant badge—*I have an athlete badge!*—and participation medal and buy braided strands of plastic metallic rainbow beads. We're ready for the opening ceremonies. I wear rainbow yarn cuffs like Wonder Woman's amazon bracelets that I knit to complement my rainbow compression knee socks and aqua Brooks Adrenaline GTS shoes. Maybe these cuffs will deflect incoming attacks. I sport a black running skirt—which I will not run in because of thigh chafe—that shows off my strong legs, and a long-sleeved orange T-shirt to cover my sunburn prone skin. My outfit is complete with the rainbow beads hanging from my neck next to my participant badge.

Andrea can't come into the stadium field with me—athletes only—so she will wait inside the Quicken arena for the procession. I am, of course, nervous walking into the participant pre-party at Progressive Field alone. The nerdy part of me outpaced the confident runner, so I stand in a long line to pay too much for a crappy cup of red wine. I don't want to need it, but my inhibitions have seeped out onto my persona making me shy Sandra. I feel more like an insider who can pretend to like dance parties when I carry my wine around the stadium, nodding and smiling to fellow athletes.

The scoreboard in the stadium blazes: "Welcome to GG9!" I observe the groups of athletes: the TEAM IORA dressed in cowboy hats and red oxford cloth shirts who will compete in the rodeo, another group in turquoise polo shirts with rainbowed TEAMCLE letters on the back, some who hold flags with their team names—Cleveland Fury. Other smaller groups of athletes drink beer and take pictures of it all with their phones and tiny cameras.

Then I see a woman in rainbow pasties and a tiny silver sequin hat pinned in her short spikey bleached hair. *Oh wow, her shorts are pasted on, too.* I track her movements. I can't find the ovaries to go and talk to her, though later in the week we will end up talking running and taking selfies together before our races. Of course, she is a runner, and each outfit she orchestrates for a road race is better than the previous one. I like how running is an occasion for which she composes fabulous outfits; dressing for running makes her a serious runner in my book.

It is time to move onto the opening ceremonies. We process into the bowels of the Quicken Loan Arena for the staging of our entrance into the opening ceremony, where I spend another hour in the line-up of *athletes* not feeling like one. I see, Ellen, a colleague from BGSU who will play tennis. We line up with TEAM RED Ohio, since we are the home athletes. There are volunteers with colored balloons that coordinate with the colored dots on our athlete badges to herd us into the procession lines. Electronic signs blaze: YELLOW LET'S GO! ORANGE LET'S GO!

"I am excited to pretend to be an athlete," I say. Ellen tells me she will play tennis, and how she has participated in the Gay Games before. I knew some of my colleagues could be here given the games are in Ohio. What is exciting about seeing Ellen is that I knew her when we were PhD students at Penn State. This feels like some kind of completed lap—running in graduate school, coming out as bisexual after graduation, running a marathon for my 30th birthday, figuring out what being a professor would mean in my 30s as a non-runner, renegotiating runner and partner identities in my early 40s—and Andrea is waiting in the stands to cheer us on. RED LET'S GO!

It is not until I walk across the stage at the very end of the procession since I'm sitting with the Ohio athletes, Ellen, and Andrea, that I feel like maybe I can be an athlete during the road races I will run. I identify as a writer and an academic and a mother and a runner, but the idea of athlete means one is competitive. I have never participated in organized sports, except intramurals in college, and I'm fairly certain that doesn't count. And then I forget about being a researcher as we watch the spectacle of the opening ceremonies: There are colored spot lights, puppets, and performers balanced on stilts; dancers twirl with tires given that Akron is the tire capital, rainbow flags unwind, The Pointer Sisters serenade, local politicians from Cleveland welcome us, and a torch of fire flares up next to the video screens, lighting our faces as one athlete races down the stairs into the arena with the Olympic torch. When President Barack Obama makes a surprise video appearance, we roar.

Monday, August 11, 2014, GG9 ~~10K~~: Postponed by Rain
Athlete in Waiting
muscles all fidget and twitch
10K not today

Tuesday, 8/12/2014, GG9 Track and Field Events, Akron, OH

Marne and I find Wally Waffles near the hotel and drown our disappointment in the postponement of the 10K by eating tricked out pancakes and waffles before the track and field events. It looks like she will have to leave in the morning before the 10K begins, so I'll be running solo. The least we can do is eat waffles smothered in honey butter and whipped cream before we cheer on other athletes at GG9 events. We meet a few women at the track and field events taking place in the Lee R. Jackson Track and Field Complex at the University of Akron, and see the only woman to run the hurdle. There was some brief rain, but the field is not as wet as after Monday's storms.

More women than men run now, but I observe more men than women here at the Gay Games. I think about Chuck's assertion on the shuttle bus about how lesbians are not serious runners because they race in the back of the pack. *Why are women in road races still seen as superfluous? And why is it that not being as fast as men means that you are not a real runner? Or if you wear a skirt? Or if you're a lesbian? What does it mean to be a serious runner? And why does someone who is not a woman runner get to decide?* I guess it depends on what you mean by winning. I haven't asked runners here about their politics, because I've focused on their experiences of running and what we love and hate about running. We've talked about finish times, but time in the context of age and ability and goals. I'm not arguing that women are more altruistic then men, but perhaps our running is relational. I hate the idea that winning at relationships makes women win at running. The 40-year-old mother of two is always mentioned as the mother of two, and her husband is always waiting at the finish line. The story is always how she got past some relational liability—mothering, partnering—in order to run fast and be seen as a _____ (fill in the blank with adjective) runner. Never do we just use the label, runner.

My internal rant is interrupted by the appearance of a news crew. Marne and I get close enough to hear what is going on. The 100-meter race is about to begin. Ida Keeling, age 99, is about to set a record. We watch her run the race in 59.8 seconds, the first person to make that time in an internationally certified race in the 95–99 year old age group.

I look up the interview later online and discover that Ida began running at 67 to cope with the loss of her two sons from drug-related homicide. Running away from arthritis and old age and running toward the finish line: Running as therapy. Running as relational therapy. Ida trains by running in the hallways of her apartment complex to the gym. She does yoga. *How cool was it to see a runner, a woman runner, make history?* I want to still be running if I live that long.

Ida Keeling is 99!
That is what you do
focus on the finish line:
"We're here to set it."

Wednesday, 8/13/2014, GG9 10K, Mustill Store Trailhead, Cuyahoga Valley, Akron

"This is like the gay Olympics." Farah tells me this on the train ride to the 10K start line. We'll start running at the Mustill Store trailhead along the towpath, the remnants of the Akron canal from the 1840s, where Sara and Joseph Mustill moved from England to Akron. I thought about driving to the start line, but I take the scenic railroad with the rest of the runners. I'm glad I did, though I've waffled all afternoon; walking would be a good warm up, but I may get lost. I'm supposed to be doing research here, but all I can think about is the race. I can talk to more runners if I take the train, but then I have to talk to people. And the 7 pm start time is throwing off my typical predawn pre-run rituals:

> I set out my race clothes the night before on the floor in the bathroom, drink a river of water, look up weather reports to determine temperature at the start and end of the race, change my mind several times about what socks and tights to wear, whether I need a windbreaker. *The compression socks with the shorter tights or the padded Thorlos with the longer tights?* I set an alarm for an hour before the time I need to leave for the start line, though I never

42 Woman, Running

need it. Two hours before the alarm goes off, I look at a clock and think about what needs to happen before the race. *Roll over and go to sleep. Roll back and look at the clock. Get up. You are not sleeping, anyway.* I get up earlier than I set the alarm—3:30 or 4:30—and pee. I wash my face, apply sunscreen and deodorant, pack a banana and a granola bar to eat right before the start, pee, put on my timer watch, pin my race number with the safety pins to the front of my shirt or pants. Curse when I stab myself with the points. Try and take a poop. Eat a piece of toast that tastes like cardboard. Pee. Put on my shoes and double-knot them. Take a poop. Remember the bra chafe from last race and slab on aquaphor jelly under my breasts and bra straps, in the creases of my legs and inner thighs. Pee. Follow the stream of runners to the start line in the still black morning. Find a porta potty and pee.

So, I'm glad I took the train. It's nice to talk to some fellow runners as a runner and not an ethnographer. I think that being a runner may be more important than being a researcher now, because I spent all afternoon deciding what to wear, trying not to drink water, peeing, and thinking about the 10K race, thinking about the 10K as a *race* and not a *run*. I will submit hotel receipts for this week as professional expenses related to research, though it seems I'm all in as a runner. The research part is proving more difficult, until I can recognize that being a runner *is* my research role this week.

As the train approaches the race start point, Farah tells me that she always wanted to be on the Olympic softball team and always wanted to run a 10K, so the gay Olympics was on her bucket list. The thing is that her MS is getting worse. "I may have to stop a lot, and it takes me 3 days to recover from a run. I will finish, but I hope to run this in an hour and a half." I don't see Farah after the race, because my goal is to run this thing in under an hour. I make sure to find her time online in the Facebook link after the race. She almost made it: 1:31:47!

I see Chuck, the runner from San Francisco I met on the bus from Akron to the Athlete Village in Cleveland, on my way to the start line. I have another runner take our picture in the late day sun; we squint our smiles into my cheap green Kodak that I brought to document the before and after. I will have to carry it when I run, though it probably won't make any difference to my finish time! We walk together to the start line.

"I am going to run in the back with the lesbians and the fat women in tutus who don't care about racing." Chuck said.

"Dykes rule!" I said. I move further up in the line-up, away from Chuckie.

What is it with the misogyny? It's been a while since I've had to deal with gay men who dislike women and find vaginas more repulsive than straight men. Earlier, when I was talking to Rene from Switzerland as we stretched on the grass outside the Metropark, he told me that *some people* say women who wear running skirts are not serious runners. Chuck bore out that attitude. Though I do not wear a run-skirt, I do not start at the front of the pack, either. I wave at him on his way to the back. *How was he reading me?* When I see Shanda, I sidle up to my dyke crush who is wearing a tutu and a feather headband. *She's not in the back.* We take a pre-run selfie together that I will post to Facebook after the run. She wears fabulous peacock feathers attached to a black headband, and I wear my good luck baseball cap with the *girls rule* badge and rainbow stripe. I haven't run a 10K since 1995 when I had the lucky 69 as my bib number. I wish that were my bib number now as the race begins.

I think about running like a dyke. I think about running like a woman. I think about running like a feminist. I think about running faster than a sexist. I think about running like a runner. I think about just running.

I just run.

I run.

I cross the finish line and have to walk circles to cool off, because I can't catch my breath. I should be talking to fellow runners, especially the woman who I paced during the last mile of the race. We finish at the exact same time. I'm having a difficult time cooling down, though, and can't really talk. I drink a bottle of water and walk. I find a fellow runner to take my picture under the GG9 banner by the finish line. We introduce ourselves, and then I take Maria's picture under the banner. I know my face is flushed pink, because I can see my blotchy post-work out arms in the crepuscular light. I confirm that when we look at the picture; I'm colored like pink marble.

"How did you do?" Maria asks.

We walk over to the results board and scan the paper printouts of the winners pinned up. When I see my name, I'm excited, and a little proud. I ran hard. "Hey, you placed!" Maria says.

Place	Overall Place	Name	Bib No	Age	5K split		Finish		Chip Time	Gun Time
					Rnk	TimePace	Rnk	TimePace		
4	195	Sandra Faulkner	129	42	4	29:02 9:22/M	4	29:38 9:34/M	58:41	58:50

As usual, I repeated a mantra to keep going at the pace I wanted. *Run Faulkner Run.* I glanced at my Garmin every quarter of a mile to see if I was running near a 9-minute mile, which is not fast for women in their early 40s, but fast for me. *Run Faulkner Run.* The top three women in my age group ran like jackrabbits compared to me; the third place winner was 7 minutes faster than me. I don't see any more names in my age group listed on the board. So there I am in last place. There was no way I could have run any faster than I did. I decide to walk back to the hotel, instead of take the shuttle bus. There are a few other runners who have the same idea, but I walk alone instead of finding some company. I make up a haiku

44 Woman, Running

and repeat it to myself as I walk back to my hotel, because it was disappointing to run hard and be in last place:

> finish line to bed
> uphill trudge but FOURTH place win!
> Fourth Place is Last Place.
>
> Fourth Place. Last Place. Fourth Place. Last Place. Fourth Place. Last Place. Fourth Place.

I check Facebook when I get back to my hotel and see that Shanda has messaged me. I had put a message about my interest in interviewing women runners on the 10K, 5K, and half marathon Facebook pages earlier. I upload our selfie picture from the 10K. We talk back and forth on the comments under the photo about running personal records.

Comment

August 13, 2014 at 9:23pm *Like*
S: Great to meet you!
August 13, 2014 at 9:25pm *Like*
SF: Hope your run was fun!
August 13, 2014 at 9:29pm *Like*
S: I just noticed your shirt. I got my PR at the Glass City Marathon a few years ago. Still trying to beat it.
August 13, 2014 at 9:29 pm *Like*
SF: I also got a PR (for my 40s) at Glass City this past May. Before I am 50, I intend to beat it.
August 13, 2014 at 11:08 pm *Like*
S: Sweet!

I look up the 10K results online after I lay out my clothes for the morning 5K. There were 14 women in the 40–44-year-old age group. I did not come in last place! They just posted the winners in the age group on the board. I almost feel like an athlete.

GG9 10K
finish line to bed
uphill trudge but FOURTH place win:
Fourth Place is Last Place.

8/14/2014, GG9 5K, Bib#1879, Cleveland Metro Park Zoo

I hate 5K races. I hate sprinting. I hate running just a few miles. I hate starting a race at a pace that makes me begin breathing with an open mouth like a dying fish. And this is what a 5K is all about. I grumble to myself as I drive from Akron to the Cleveland Zoo in the pre-dawn. *I just ran 12 hours ago, and I have to run another race? I'm running 13.1 miles on Saturday?* I see runners from last night as we stretch pre-run. We talk about seeing this race as fun, because many of us are thinking about the half marathon on Saturday. *Yes*, I will focus on this as a fun run. *It will be fun to run around the zoo and see animals.* I look around before we line up and try to spot the serious 5K runners. The ones who look like runners. The ones not dressed in tutus and running skirts. Shanda and I take another pre-run selfie. Once again, I admire her run outfit; the white feather puff attached to her headband is fun.

"I thought you said you were slow? You placed in your age group," she says. I'm glad she was following me. We line up behind the faster runners at the start line together, and I run ahead of her, though she will finish the race 30 seconds before me.

I deserve to suffer during this run, because of my crappy thoughts about *real 5K* runners. As if real runners don't see the race as fun. As if fun and race are mutually exclusive terms. As if those misogynistic attitudes about women and running have any merit.

GG9 5K (chip time 30:31, pace 9:51 mile, 6/11 women 40–44 years old)
12 hours after 10K run past a sunning camel
twice, two hills, two men in tutus
two shirts that read, "tutus make you run faster."

Saturday, August 16, 2014, Bib#2088, GG9 Half Marathon, Age 42, Akron, Ohio

Miles 1–4

My legs are stiff, and I'm having a hard time warming up. I have to pee. *Thirteen miles is a long way.* I see the 10:18 minute/mile pacer. *How can someone hold a sign up for that many miles? Maybe I can break my 2:15:00 half marathon barrier.* I follow the pacer. I still have to pee. *Didn't I pee three times before the start? If I stop, I will not keep up this pace. Okay, I will run this race in 5 miles increments. That will make it seem shorter. I can run for 5 miles.* I hear another runner huffing behind me.

"A hill? I can do this," he mutters.

"You can do it," I say.

"I can't believe we have to run another 10 miles," he says. Running has slapped this man down and won. "This is my second half marathon, but I'm not sure I'll make it." He wears the face of a defeated runner.

I encourage him up the hill and continue on. I need the sprinkling of spectators who yell for us, and this runner needs me to keep running like I mean it. "Nice Work, Runners!"

46 Woman, Running

Mile 4.5

I see a porta potty on the top of the crest. *Ah ...* When I stop in front of the blue plastic door, I see the lock. *Are you fucking kidding?* I have to pee. I walk behind the locked john, squat beside a bush, pull down my running shorts, and pee. *I don't care if anyone sees my ass.*

As I run out from behind the potty, I see a male runner discover the padlocked door. "Are you kidding? This is padlocked?" he asks the volunteer, who has not noticed. He clutches his bottom, so I figure he has to do more than pee.

I have lost the 10:18 pacer.

Miles 5–7

After my pee stop, I see the 10:40 pace group and follow them. I'm feeling good. *Maybe just maybe I can run this thing.* As I pace into the pack, I see a sign pinned to the back of one pacer's shirt: "Thank the volunteers." As I run beside them, I wonder at their steady gait and intuitive timing. *They must have internal racing clocks. Do they just feel the pace in their legs? How can I learn to do that?*

"How do you keep pace? Do you keep the pace in your head?" I ask. I'm disappointed when one pacer answers, "Gadgets." *So there is no pacing super power.*

As we run by police officers who keep the race course clear, and volunteers who tell us which direction to run, many of us do thank them.

"Thank you for being here."

"Thank you for cheering."

"Thank you for your service."

"Good morning," I say to many spectators. They often talk back. "I like your socks!" My rainbow compression socks are magic, though they won't help me keep up with the 10:40 pacers. I stop short when a car tries to plow through the race route in front of me. A police officer stops the woman who rolls down her car window to ask how she is going to get to her house. She does not cheer for me as I run around her bumper.

Some people smoke on their porches and stare as we run by. Some wave, but there isn't as much energy and crowd enthusiasm as in the larger races I've run in Detroit and Cincinnati. I have to go to the bathroom again. This time the porta john is not locked. I lose the 10:40 pacers.

Mile 7.5–10.5

I have run 1:43:00 minutes according to my Garmin. *Sigh.* The good running feeling is gone. I see Chuck. He crosses over from the left side of the street to talk with me. He runs with me only long enough to say hello, because he's feeling the race and is making good time. He waves as he runs ahead with another runner from his running club. *I can't believe I still have so many miles to run.* My legs feel like they are made of granite.

I say hello to another runner as I approach beside him. *Maybe I can distract myself.* "What are you running?" he asks.

"I'm doing the half marathon," I say.

"I'm doing the full marathon, and I'm taking my time," he says. David, from Boston, tells me I can run ahead. He is just pacing himself and not in any hurry.

"How many marathons have you run?" I ask.

"Fifteen!" he said.

"Wow. I've only run one marathon. I ran all of the road races here, so I'm tired. The thought of just running a half today is hard." I slow down to run beside him. We talk about living in the moment, the difficulty of children. He is father to six with his partner. They fostered, and then adopted them all. I tell him that I have just one child, a 5-year-old daughter.

We talk about the Department of Social Services and how he thinks that some services are pushing kids to come out. "I'm worried about HIV. My 21-year-old son is having all kinds of sex." David tells me that the social services people are irritated with him because he pushes for services. "I do it for my kids and not for me." He tells me he should get support because he takes in the retarded and the troubled.

We pass students at the University of Akron, and David yells, "Way to go runners!" The students were talking to one another and not cheering us on. "We are running, and they are thinking about what they are going to have for breakfast. Get with it!" David says he runs so that he doesn't weigh 300 pounds and so that men would want to have sex with him. At the Gay Games in Cologne he had sex with all kinds of men.

"*I didn't need to know that.*" I think. "*Why is it all sex and party time here?*"

"I wouldn't take my kids to Pride or the Gay Games because that is not their identity. Parents shouldn't push their identity on their children. They shouldn't live their lives through them. One of my six kids is gay," he says. *I consider how one should talk to their children about identity. I should figure this out given I study this stuff.* I tell him that my daughter and partner are here and will cheer me on around mile eleven. *I brought them here, so does this mean I'm pushing my identity on my daughter?*

"My 21-year-old was taken to a gay bar when he was 17 by someone on the staff at the services. The young gays are being pushed out of the closet too soon before they know anything," David says. I'm listening to him, amazed that anyone wants to parent that many kids. One is doing me in. I've only recently integrated the mother role into my list of identities.

David continues. "Things were different for me. It was more difficult growing up today." He was certain if he worked hard, he would have a job, and things would work out. Not so with kids today.

"You're right," I agreed. We talk about being in the moment running, feeling good. I tell him about my father and his aneurysm, how I'm grateful that I can run.

48 Woman, Running

"Did you hear about the case of the woman who transitioned to a man and had a kid? What about the kid?" he asks. I nod. "And then he sued the fertility company. What a rotten thing to do. And selfish for the kid."

"Do you think that parenting is tied to biology?" I ask.

"No. I always wanted to be a parent. I've been a parent for 25 years," he answers.

"Why is it that we have all of these non-gender conforming people? What is this? I think that people should conform," he says.

"I'm director of Women's, Gender, and Sexuality Studies at my university. I think things are complicated," I say.

"You're the expert," he says.

"I think gender is more contested than our biological sex. That is what makes it hard for us to accept. We are so used to things being masculine *or* feminine," I say, and then I change the topic. *Why is it that I keep having long conversations with men and not women at the Gay Games?*

"Did you carry her or did your partner?" he asks.

"I did, because my spouse couldn't. Though, I wish he could have!" I joked. I have come out to him as having a male partner.

"You shouldn't use the term partner. It is confusing to us homosexuals, and the games are full of homosexuals," he says.

"I'm one of the homosexuals! I'm bisexual. I just happen to be partnered with a man," I say. *Who are you? The Gay Police?* I didn't ask this question.

I use the term partner, because of this very attitude. I'm not straight. And even if I were, I should be allowed to use a term that best expresses my perception of my relationship. His admonition reminds me of my friend JD's ignorant views when I lived in North Carolina. JD told me he didn't believe in bisexuality, that someone who identified as bisexual was just not being honest about being gay. We would have dinner together, cruise men, and pontificate about attraction. Over Cesar Salad with blue cheese and fresh squeezed lemon wedges, I talked about men and women I found attractive. He dubbed me *semi* for semi-straight. My victory occurred after a year of such meals, on the day I moved back to Pennsylvania, when JD hugged me goodbye and said, "You are bisexual!"

After we have talked for a few minutes longer, David says "Don't let me hold you up." I'm tired of talking to David, so I say, "I'm going to just get this race over with. Good luck!"

"You go girl," David says.

I turn around and say I prefer woman. "That doesn't have the same ring, though. '*You go woman*' sounds like GO make me a sandwich woman." I laugh as I turn back around and run away. I speed up with my power of sheer will, thankful that my penchant for stubbornness is useful at times. Getting away from this Queen is good motivation to pick up my tired legs. I could have run faster without the 3-mile conversation, but it diverted attention away from my breaking body. I jinxed myself at the start line by thinking about 13 miles as a long way.

Now I'm back to being tired and having this run be all effort. Three more miles of being in my head after a hard week won't help, either.

Mile 11

I hear the slap of another runner's shoes on the pavement. The pounding distracts me. I'm usually in my runner zone by this point in the race and able to ignore everything. *Ugh, that heavy breathing and plodding footfall is super annoying.* I want to run away, but I can't seem to move any faster. I'm sure I'm gasping like a dying fish, too. I should see Josh and Mimi soon. It's strange that they will be cheering for me here and waiting for me at the finish line. I usually run races solo or with Marne, and we drive ourselves home. At the end of this race, I will be *MOM*, not Sandra, *the runner*.

I round a bend, and see Mimi holding up the sign she made the night before.

RUN
MOMMY
RUN!!

She doesn't yell for me, and hides behind her sign. Josh tries to coax her out. I stop for a moment to high five her hand for power. I needed to see them.

Mile 12.5

As I turn a corner to head into the final leg of this wretched race, I see a woman pacing me. I should push forward and pass her over the finish line, but I can't make myself go. *I need this run to be over.*

Mile 13.1

"Crossing the finish line, Sandra Faulkner from Bowling Green, Ohio."

I have never heard my name called before during a race. I'm disappointed in my time—02:23:23 and a 10:56 mile—but I got my finisher medal. I wasn't sure I would. Three road races in 78 hours were more than exhausting. I remember the reporter at the 10K who told me that not every race can be a PR. I hear Mimi and Josh yelling for me. When I look at the picture that Josh took of me crossing the finish line later, I notice my feet plastered to the pavement, the droop in my posture. The smell of defeat clings to my sun shirt when the woman, who I do not pass, comes up to me.

"I was pacing you the last few miles," she says.

"How did you do?" I ask.

"I did my best ever. I got a PR!" she says. We shake hands.

"Congratulations!" I say. I find out later that if I had passed her, I would have been fourth instead of fifth in my age group.

50 Woman, Running

I see Mimi and Josh by the snack table. There are peaches! The volunteer who hands me a peach wears a circular rainbow cowl. I ask her for another peach, because Mimi loves them, too. "Did you knit that?" I ask. "It's beautiful."

"You're a knitter and a runner," she states. There is no need to ask the question.

"Of course! Knitters and runners are good people," I say.

"Are those peaches from Georgia?" I ask.

"Are you a Georgia peach?" she asks me. I swear she winks.

Summer Run
sunscreen, sweat, urine
iPod drops in the toilet
dive in thereafter?

12, October 2014, München 10K, Age 42, 00:57:45, 9:17 mile pace, 72/209 Women 40–44

Run the Rhine

Do not begin by the too green Lindenhof
with the stony stare of Princess Stephanie,
the Mannheimer dogs off-leash who get to pee
where they please while you must hold it in.

Start at the urine-soaked graffiti-sprayed
tunnels under the tracks,
do not flinch with the sound of the clack
as the Deutsche Bahn thunders over your head,

dodge the post weekend pile of vomit
carnage of used wrappers and bottles
until you cruise out of the fragrant disorder
into the Schlosspark thick with rabbits

like Watership Down, skitter past the murder of crows
that eye you with *the eye*, you stranger,
they were here first and are hungry,
can outcall the flocks of whiny lime green parakeets.

Have keine Angst, no one will talk to you
as you hear only the flap of your hat
smell the proof that Germans like dogs
better than runners and children.

Stumble over chestnuts, the grit from the Promenade,
into the Waldpark with only suggestions
of the city traffic and sirens, the endless construction.
Do not smell the stench of Ludwigshafen

over the bridge when the wind blows just so.
Try and get lost in this pretend forest
that is better than your dream of Germany,
know there will be no hidden trees for a WC trip.

Decide that public urination is fine and fun
like breaking some half-remembered grammar rule.
Pull up your pants and turn around by the snake-neck bend,
you can chase the cargo ships stacked with stuff.

Laugh when der Schlauch by the spectator benches
elbows them sideways with a strong arm,
so you can not cry this rotten nostalgia into the Rhine
as you limp your leave.

My entire sabbatical semester in Mannheim, Germany, I've run injured. From August 19, 2014 to December 2, 2014, I live, work, and run in Mannheim, the eighth largest metropolitan region in Germany, located at the confluence of the Rhine and Neckar Rivers in the northwestern corner of the state of Baden-Württemberg, (re)learning the German language after a 21-year hiatus. My feet hurt even when I'm not on them, so the miles and miles Josh, Mimi, and I walk around southern Germany keep my plantar fasciitis looping on a closed track of inflammation. I hopefully buy a pair of Birkenstocks, but even German engineering can't fix my feet. I'll hobble around until January when I go to the podiatrist in Bowling Green, who sees my feet from across the room and tells me how swollen my plantar fascia are, even without a close examination. I only visit the doctor because a colleague has asked me why I was limping.

I enroll in a German as a foreign language class taught by two teachers from a high school in Mannheim and spend Monday and Wednesday nights immersed in culture, grammar, and formal language instruction. I teach a 7-week course on Gender and Interpersonal Communication in English at the University of Mannheim, where I'm a visiting scholar. Most weekends, Josh, Mimi, and I travel by train through Germany. I run along the Rhine River three times a week and watch the river push barges full of coal, cars, and other cargo. I spend time with a new German friend over wine, coffee, and apple cake, practicing the art of German conversation. My 22-year-old self is living the life I dreamed of when I fantasized about quitting graduate school and moving to Germany after a run of German boyfriends. The reality of the middle-aged version of my German adventure with American child and American spouse takes some adjustment.

52 Woman, Running

I sign up for the Munich 10K and not the half marathon, because the Gay Games wore me out. I'm not sure how many miles I will put in during my research semester, but I want to run a road race in Germany. On any given day, I know I can run 6 miles without any intense training. Running in Munich appeals to me, especially because the race begins and ends in the Olympic Stadium. I only remember the Rathaus-Glockenspiel in Marian Platz when I was backpacking around in 1991. Josh, Mimi, and I make plans for a long weekend trip to Munich in October. We will meet my friend Gerhard, from my undergraduate days, at the Hofbrauhaus one afternoon. Then I will run the race on the day we train back to Mannheim.

I trick myself into running a PR at the Munich race. During my non-training runs through the park by the university, and along the Rhine, I mutter to myself that I only have to run the 6.2 miles in under an hour. I'm on sabbatical; I should be able to relax and focus on writing and not these other non-academic goals. I usually run around 3 miles at a time, so I try not to think that running 6 miles in under an hour may be a stretch. *If I can run the 10K in under an hour, then I won't be an embarrassment.* I switch my Garmin to track kilometers, but can never get the feel of how they translate into miles. I switch my Garmin back to miles.

I take the train to the stadium from a stop near our hotel. Mimi and Josh will meet me back at the hotel after the race. They won't come with me to watch, though they will see other half and full marathoners run around the city. The 10K race course is too far from our hotel. We have no cell phone, so we are making plans old school. I guess how long it will take me to run, shower at the stadium, and return to our meeting point. After I find the spot to drop off my post-run clothes in the pink plastic draw-string bag under the stadium bleachers, arranged by race number, I go and sit in the stadium seats. I see the finish line at the bottom of the stadium stairs on the track.

I am in Munich in the Olympic Stadium! I will get to run a lap around that track like a real athlete!

The conversations I overhear about running times before the race are familiar to me, as is the last minute stretching, eating of bananas and energy gel, the nervous pacing, the pinning of race numbers to shirts. I take a picture of my banana, good luck rainbow socks, and race number as I stretch on a piece of pavement under the Olympic bleachers on my way to the start line.

The difference is the accented English or what I dub Deutlisch—half English/ half German, the language in which I am fluent. "Start. Eine Minute." One minute to race time. I can't get my Garmin to find a satellite, so I will just use the timer function. The day is sunny, so I'm glad I have my ball cap with the sexy flap of fabric in back to keep the sun off of my neck. We are starting at 11 am, which means the course will remain full of sunshine. I smell like sunscreen and nervous sweat; the stinky exertion sweat will happen after a few miles of racing. When I step onto the spongy black pad under the start sign that activates the timing chip I tied to my shoelace, I go full throttle. "BMW, naturlich." I hear part of the announcement. *Of course BMW is a sponsor!*

Once we are past the start line, there are fewer signs along the race route compared with the US races I have run.

Ihr Seid Super, Haltet Durch (You are super! Keep going!)

Some spectators wave the blue checkered Bavarian flags. A few take photos of passing runners. The roads where we run are so narrow, it is like we are spawning salmon. I keep touching other runners, because I'm boxed in by elbows in motion. Usually I am able to carve out a space after the first mile, but there are too many of us on these slim roadways. I weave around runners when I can, trying to keep 10-minute miles. But it is difficult to gauge my pace from the timer on my watch, because the marker signs are all in kilometers. *Duh, Sandra.* You've been living in Germany for 2 months and still feel happy when you hear German, though the math conversions are impossible to calculate while running.

I pass the 3-kilometer sign and speed up. I try and enjoy the sites we are passing, and ignore the feeling of drowning in a sea of runners. I'm hopeful that I will make my goal, and every chance I get to pass another runner, I take. Then, we enter the stadium for the final lap to the Ziel. I pass as many runners as I can on my way to the finish. I hear the announcer say something about "Dieses Wochende."

This weekend indeed. When I stop my watch after I cross the finish line and get my medal, I see that I ran almost a minute faster than my Gay Games 10K! I go to the keg of hefeweizen and get a full cup of beer—only the second one during my time in Germany, because I'm not a beer drinker. I have been drinking the trocken Rieslings produced in the Rheingau and Rheinhessen, but there is no way I cannot drink a cup of Munich beer after a victorious lap in the Olympic stadium. I also get a Brezel and a cup of water to rehydrate. *Ich liebe es zu laufen. I love running.*

PR: Personal Record
you become body
you become your body you
become your body

July 2015, 3 miles, Age 43, Old San Juan, Puerto Rico

This summer, the summer of my 10-year wedding anniversary, a colleague of mine and I interrogate interview transcripts of new brides' talk about planning their weddings. We burst out of the usual interpersonal theory suit and dress our analysis in Leslie Baxter's relational dialectics—one of the few critical interpersonal theories—and unsnap the narrative that the wedding day is the most important day in a woman's life. Women are supposed to fall into a culturally approved narcissism and take the role of bridezilla; "It's my day, and I'll do what I want to." The woman in expensive lace dances over the couple, the marriage, escalates to the top of the stage.

Marriage, like other relationships, is all change, but we perpetuate the dominant Hallmark Channel image of marriage as continuous romance. This rainbow romance can't exist, like a rainbow's ephemeral existence; no relationship can bear that impossible weight. I wonder if my relationship can bear the weight of critical analysis and personal writing? I want to show what being in this relationship, my marriage, feels like, and the theories about personal relationships I fought in graduate school can't show my non-engagement, the reworking of the supremacy of marriage as stifling institution to living body with bruises, stretch marks, and stitches.

But before I get too smug about my cheap wedding—I scotch-taped the receipts for the clothing into an album—and my embrace of the anti-wedding and focus on marriage, I consider that Josh and I had not one, but three celebrations—the legal marriage, the love marriage in Madrid's Parque del Retiro, and a party for 50 at Sheldrake Vineyard off Cayuga Lake a year after the legal marriage. We had talked about not wanting to spend money on a wedding because we were in our 30s and self-sufficient. We owned a house, a car, had "real" jobs. My parents had paid for my undergraduate degree, which was better than one day and one celebration.

I think about marriage as I run along the Paseo del Morro in Old San Juan, choking on the cat smells in the stinking hot sun, while Josh sleeps in our hotel room. This is our vacation sans child to celebrate 10 years of being married. But, I do not need a vacation *for* my marriage. I do not need a vacation *from* my marriage. I self-medicate with marriage and all of the tasks it takes, accept my impatience and anxiety about impossible expectations as the only permanence. Marriage is not a solid state despite the platitudes we circulate: *You're the same person I married X years ago. She hasn't changed in X years. He hasn't changed in X years.*

X is not a constant state.

City Run
graffiti pisses
catch laces on cobblestone
no one notices

9/5/2015, Boy Scout Half Marathon, PR (Personal Record), Bowling Green, Ohio

Place	Net Time	Sex/Total	Gun Time	Pace	Name	Age	Sex	City
212/337	**2:11:42**	66/152	2:11:56	10:05	FAULKNER SANDRA	43	F	Bowling Green

I see Steve before the race. "What are you hoping to run?" I ask. The ubiquitous question we runners ask one another before a race. We shake out the kinks in our legs as we talk by the porta johns near the start line on the BG fairgrounds. I will visit them a few more times before the start; my bladder has pre-run nerves. Other runners sprint back and forth in front of us, some stretch out in the patch of grass by some parked cars. We all engage in our pre-race rituals. I kick my legs in front of me, high step march, and whip my knees around like a frog to stretch my hip flexors. Steve eats a pack of GU energy gel. I finish a banana. I realize that I often look for Steve at these local races, so I must be a part of the running community, though I'm not an official member of any running club or group.

"I'm training for the Columbus Marathon, so I need to run well," Steve says. I met Steve on a long run around town last year. I noticed this middle-aged man was pacing me as I turned onto Poe Road. We said hello. We were running at the same pace, so we started talking and running together. It turned out that he was married to a colleague of mine who ran the Women's Center. We talked about gender and work and running. I ended up running 8 miles with him, which was not in my training plan. Steve offered me a glass of water at his house, and drove me back to my place. We then saw one another on runs around town and at local races.

"I haven't run much, because my feet hurt. My plantar fasciitis made me take a month off in the spring, so I'm just running to have fun. I will run how I feel." I have a goal to run a 2:10:00 half marathon sometime in my 40s, but I know that today is not going to be that race. I have not trained, and I'm coming off an injury—my feet, always my feet.

What I felt was a PR (personal record), and I almost made my goal! I do not look at my watch during the race. I remember the last time I ran the Boy Scout half marathon dehydrated, and I'm ecstatic in my running body today. It is not as humid, and right when I feel the heat flush my white cheeks in blotchy streaks, it starts to rain. I inhale that moist air, and still do not look at my watch. Running in the rain on a late summer day around my usual routes, I am all runner. When I cross the finish line with my runner high, satisfied with how I feel, I find out that I ran 4 minutes off my best half marathon time.

What If
we run by feel like
what we feel is what we run
like run full of feel

July 17, 2016, Injured, 2 Miles, Age 44

I wait until Josh and Mimi leave for service at the Unitarian Universalist Congregation before I lace up. I tell them I want to relax at home. I have decided that I am going to run with my arm in the sling without any

56 Woman, Running

witnesses. Working out on the elliptical machine at the gym is just not the same. I know that I'm still a runner, though it has been 3 weeks since my last run. When I walk down the sidewalk carrying my achy shoulder in that sling and pass runners, I'm wistful. I want to be running. It is painful for me to see people out running when I can't.

I eat a chocolate raspberry cookie for fortitude, though all of the running websites argue that a Medjool date energy ball would be the best choice for some carbohydrates. I take my chicken wing arm fastened in the sling with my "screwed" shoulder limping behind. I run a few practice steps. It's like bouncing along with a pack full of rocks, my center of gravity thrown ahead of me. I imagine I look like I'm late for a meeting: black sling, button up shirt, and shorts hurtling along the sidewalk. Desperation is a terrible running companion. I manage two miserable miles.

Always Running
sidewalk full of late
running even when you walk
a good pace for life?

August 31, 2016, 2 Month Post-Op Visit with Dr. Gomez, Age 44, Wood County Orthopedists

Doc: How are you feeling?

Me: My shoulder hurts, but I almost have full range of motion back. [Demonstrates by folding left arm behind back, palm flat against the small of the back.]

Doc: When does it hurt?

Me: After PT and after running. Icing helps.

Doc: You know if someone tells me, "Doc, my arm hurts when I pull of this Band-Aid," I tell them to not pull off the band-aid. You could give up running. You can do the stationary bike, the elliptical machine-

Me: That is boring! Running is weight bearing, good for my bones. I know I am arguing with you. [Continues arguing.]

Doc: I have a patient who was a runner. She is just 50 years old. I replaced both her hips. I'm going to replace her knee next month.

Me: [I know I am playing into that cliché of runners as mad maniacs, but I think I've proven that runners are strong. Or he is giving me some other warning?]

I have a goal to run a 2:10:00 half marathon in my 40s. I turn 45 in 2 months.

Doc: When is your next race?

Me: Saturday. [I won?!?]

Interlude: I go to PT after the visit and recount the conversation to my physical therapist as she manipulates my shoulder in range of motion exercises.

PT: Orthopedists always tell their patients not to run. They think running ruins joints.

Me: Of course, running is how I jacked my shoulder in the first place.

PT: I run with my plantar fasciitis.

Me: That, too. I feel so much better when I am running. It keeps me healthy. And I know when to take time off. [This is a lie I tell myself again and again as I am in PT, *because* I run through injuries. My athlete friends tell me this is what athletes do: They run when it hurts.]

PT: Do you want to ice your shoulder?

Me: I will ice it later after I go for a run.

Body Is Mind

I am all body
aching arthritic hip feet
keep run-run-running

9/3/2016, Boy Scout Half Marathon, 2:15:59, 51.1% Age Grade, Women 40–44, BG

The last 3 miles I am all body: the aching right hip joint, the arthritic feet, the jacked-up shoulder with two stainless steel screws, the left ACL crafted from a piece of my own hamstring and held into place with surgical thread and two more screws, the muscles that are burning oxygen at maximum capacity. I turn the corner at Conneaut Avenue onto Fairview Avenue and the last leg of the race. I *am* runner. This is my 13th half marathon.

"GO, Sandra!"

I hear my name and the clapping and yelling as if I am running under water with cotton balls shoved into my ears. Anything not part of my legs and lungs moving body forward is superfluous. I recognize my neighbor in some part of the brain that is not body. *Oh, that is Sara from down the street.* I am so tired. Then, I am past thinking of a post-run nap. Past thinking about all of the aches. Past thinking period. This is the run after major shoulder surgery that knocked me off the training horse. I cannot run any faster in this moment. *Keep running.* All I can think and do is run. My body *is* running: MINDBODYMINDBODY. Running is embodiment. Running is the mind; the irony is that running is bodily mind, and what holds one back is the mind. This mantra plays inside my head: *Just run like a horse. One, two, three. One, two, three.* Counting breaths and being runner. *Keep running like a horse. Keep running like a horse.*

You Are

if you run, you are
see yourself as a runner
athlete woman run

Bibliography

Baxter, L. (2011). *Voicing relationships: A dialogic perspective.* Thousand Oaks, CA: Sage.

Faulkner, S. L. (2014). *Family stories, poetry, and women's work: Knit four, frog one (poems).* Rotterdam: Sense Publishers.

Jutel, A. (2009). Running like a girl: Women's running books and the paradox of tradition. *Journal of Popular Culture, 42*(6), 1004–1022. doi:10.1111/j.1540-5931.2009.00719.x

Schleis, P. (2014, August 12). Great-great-grandma from New York City sets record at Gay Games 9. *Akron Beacon Journal.* Retrieved from www.ohio.com/news/local/great-great-grandma-from-new-york-city-sets-record-at-gay-games-9-1.513047

Schnohr, P., O'Keefe, J. H., Marott, J. L., Lange, P., & Jensen, G. B. (2015). Dose of jogging and long-term mortality: The Copenhagen City Heart Study. *Journal of the American College of Cardiology, 65*(5), 411–419. doi:10.1016/j.jacc.2014.11.023

3

REAL WOMEN RUN

I hate running. The first mile is the hardest.
I love running, but I hate running. (Sabrina)

Running is like fighting against myself and winning. (Mavis)

What I enjoy most about running is it is a sport that regardless of age, gender, ability, race, and reason for doing it, no one is judging you. The only competition you have is with yourself. (Jaclyn)

Why Do Women Run?

Women experienced running as social and solitary, pleasurable and painful, dangerous and empowering. *Real Women Run* is about women running; about identities in motion, the inseparable mind–body connection, and running as solitude, physical and emotional strength, and community. I wanted to know: *What are women's embodied experiences of running and being a runner?* I talked with 41 women via phone, IM, email, Skype, and in person about their running experiences: their perceptions of their running bodies, what running means to them, and how running fits into their lives and social networks (see the Appendix at the end of the book for questions and demographic information). Women were mostly White (36), but not all. Mostly straight (37), but not all; some had kids (19). They represented a variety of backgrounds from academic to artist to army wife to biologist to caterer to copywriter to nurse to student, and their ages ranged from 25 to 56 years old.

60 Real Women Run

"Why do you run?" I asked.

I run   because I'm competitive
              because I am a Marathon Maniac
                    to release stress
             to meditate
      to feel strong.

     I ran   because I was in Africa
                to have some control
                    to control chronic pain
                        to lose weight
                for my health and my kids
             when my husband was deployed.

           I run   to remember
                    to remember I can do hard things
                  to forget
                to get over this divorce
            to run away from family violence.

     I run   to change my life
              to keep centered
                for psychological health
          so I can drink
      so I can eat
  so I can eat cheese.

    I run   to get rid of marriage weight
              and baby weight, too.

     I run
       for my good
          because I hate it
             for fun
       to keep a schedule.

           I run   as punishment
                      as connection
             to become a warrior.

    I run
       to feel strong

to ease depression
to help endometriosis
to see other women
because of misogyny.

I run
because running is mine.

I run.
Running is mine.

I run   to be free
to do my own thing
for me time.

Runners don't slack.
Running is cheap.
Running is efficient.

I run   because that is who I am. (41 Women Runners)

Women who run: women with disabilities, fat woman, women who've recovered from physical injuries, trans women, migrant women, Indigenous women, depressed women, women with no time, women with kids, ladies of leisure, school-girls, retirees, mothers, aunts, grandmothers, queer women, straight women, slow women. Scrutinize any of these categories and a set of stories that defy generalizations will emerge, stories that destabilize the big stupid myths that say women can't run, that only certain kinds of women can run, that's it's too dangerous, that it's too unfeminine, that it's a sign of trouble. (Menzies-Pike, 2017, p. 232).

Running for the women I talked with was about being in control, being blessed, being healthy, setting goals and meeting them, challenging the self and body, being strong, being safe, and for some, part of a spiritual practice. Running was part of their identities and a way of life. They described good running as effortless and for many runners, there was no such thing as a bad run, because any time you ran, you accomplished something. Bad runs, though, did occur, and many would wait for the one good run that redeemed their running practice. Women ran to feel better about themselves and to run away from selves they didn't want to be; they ran toward better selves, relationships, and ways of being. The majority of women described "real runners" aka professional runners as athletic, muscled, thin, lean, and not big-chested, even if that is not how they perceived themselves.

62 Real Women Run

Why Do Women Begin Running?

"Tell me how you began to run." I asked.

I started running.

I started running in grade school, in my teens, in eighth grade
      when they called me chubby in high school.

I ran in high school for basketball, for track, for cross-country, too.
      [I hated running in high school!]

               I played soccer in high school, and had to run
               around the soccer field in practice
               and up/down my road a few times.

I started to run when I gained weight @26
      when I was 32 and overweight.

               I had always wanted to run a marathon.
               I was 32 years old. I started walking
                    on my treadmill every night while I watched TV.
               Before I knew it, I started jogging.
               As my weight dropped, it was easier and easier.

               Then spring came and I started to run outside. It was exciting
               to reach each new milestone:
                    my first 3-mile run
                    my first 5-mile run
                    my first chafing from my sports bra
                    my first bad blister
                    my first "pee in the woods" because I couldn't
                      make it home
                    my first "under 10-minute mile."

               It was pure adrenaline. After running my first 5-miler,
               I signed up for a full marathon.

I began to run after college
      in graduate school, in 1999, in 2006.

I started to run again
      when running stopped being a punishment.

I started doing serious training @40.
        It was New Year's Day 2000
        that date simply begged
        for a BIG new year's resolution.
        I planned to lose 50 pounds
           and start running.

I began to run when I was 48.
        At first, I just ran tiny distances ...
        a tenth of a mile a day the first week.
        I worked on my own for the first year,
        working up to 5 miles, 5 times a week.

I run off
    and on.

I took a year off when I broke my ankle.
I started to run after a bad relationship, the divorce,
      after my first child, after my fourth baby.

        After three children and too much home cooking,
        I found myself 85 pounds too heavy.
        I started Weight Watchers
        and bought a treadmill.

I began to run with my daughter
      with my husband, with my father.

I started running to get off meds,
      because of depression and alcoholism.

        I started when my first marriage was unhappy.
        I was addicted, and it was unhealthy.
        I loved it:
            I would run 6 miles 4 days a week.
        I married a high school boy, and I was young.
        I was Happy.
          Unhappy.
             Happy.

I started to run because of trauma,
      after my son was killed.

I ran in the army because it was expected,
    when I was deployed in 2010.

I started running after I organized an event for injured vets:
    If they can be physically active
    from wheelchairs,
    or with missing limbs,
    I have no excuse.

I ran my first marathon in 2007,
    my first 5K in 2013.

I ran after a couch to 5K.
I started running to find time for myself,
    to do what that woman on the treadmill could do.

I started running for something to do.
I ran to quit smoking
    when I did cross-fit in 2011.

I kept running
    when it was better than walking. (41 Women Runners)

June 14, 2014: Thoughts After Talking to Women Who Run

Runners run.
Women runners face safety issues.
Running is good for body love.
Running is my happy and helps me appreciate MY body
    and its strength and what it can do.
Running makes me a better mother. Running helps me deal with kids.
Running is different for mothers than women without children. (Field
Notes)

When I analyzed women's stories about running using poetic inquiry (see
Chapter 5 for more details), I found the following themes: *running as health*; *running as accountability*; *running as relational practice*; *running as safety and danger*; *running for a reason*; *running with your body*; *running as expansion*; and *running as self-definition*.
As I was writing this chapter, I listened to women's interviews on my iPod during my own runs, running in stride with them as a running buddy. What follows
is women runner's stories of running, not running, running toward and running
away, and identity and health negotiation.

Running as Health

> When I'm asked what running means to me, I tell people that running saved my life, because it has. (Teota)

Every woman I talked with told me that running was vital for their mental and physical health. Running is popular because it costs little, can be done anywhere, and has significant health benefits (Szabo & Ábrahám, 2013). "All you need is a pair of shoes and a running route" (Emily). Running allows women to maintain a healthy lifestyle and provides a means of escape from worry and anxiety (Shipway & Holloway, 2010). Most recreational runners, like Pam below, run to improve their health (Shipway & Holloway, 2010).

"I Call Myself a Stubborn Turtle," Pam Said

I just completed my fifth Marine Corps Marathon.

It's changed my health—both physical and mental—immensely. I was on high blood pressure medication, and was overweight headed to obese. I'm now at a healthier BMI but more importantly, my mental health improved.

If I'm having a bad day, I just go for a "run" and clear my head. (Seriously, if I break a 13-minute mile I feel like a rock star!)

Running Is the Best Thing That Could Have Happened to Me

Consider research that shows the risks of cardiovascular disease and other deaths are lower for runners compared to non-runners (Lee et al., 2014). Runners actually add years—approximately 3 to 7—to their lives by running, even if they are slow, overweight, smoke, drink, or run sporadically (Chakravarty, Hubert, Lingala, & Fries, 2008; Lee, Brellenthin, Thompson, Sui, & Lavie, 2017). Even more interesting is that the extra years were tied specifically to running and no other forms of exercise. Just 1 hour of running can add 7 hours to your life (Lee et al., 2017). Jacyln liked running because "running is a life-time sport." Indeed, consistently running through one's life can slow aging and improve quality of life by delaying disability and potentially preventing dementia; many runners feel younger than their chronological age (Chakravarty et al., 2008). Alex worked on an army base and ran a *Centering Pregnancy* group where she focused on the benefits of running and exercise as a way for women to be healthy in the pre- and post-natal period.

Half of the women I talked with began to run to lose weight and to help with difficult and bad relationships. Most women continued to run because it brought more positive things to their lives. Janelle wrote me an email that is indicative of how women who begin running to lose weight continue to run because of the joy running brings:

> I heard through a friend that you are looking to talk with women runners. I am a 32-year-old mother of 7 who DESPISED running while I was young and used to forge doctors' notes and tell my young male P.E. Teacher that I was on my period in order to get out of any sort of running in High School.
>
> After my fourth baby, I started running as a last resort to lose some weight and fell in love with it and have since become completely addicted to running. I'm registered to run my fourth full marathon this fall and have run three half marathons as well.

Running also brings some health downsides; runners are at higher risk of osteoarthritis and cardiovascular malfunctions and running-related injuries such as tendonitis. Danielle said, "Runners are masochists," because when one works hard, running hurts. She did say that the pain should be a tolerable pain, or at least that is what she tells her clients as a personal trainer. Whether one suffers these running-related injuries, though, depends on factors like age, prior injury, body mass index (BMI), running frequency, and the condition of one's running shoes (Buist et al., 2008). Women face different physical challenges as runners than men. For instance, there are marked changes due to hormones, aging, and pregnancy that influence locomotion, impact, and injury (Dufek, Mercer, Teramoto, Mangus, & Freedman, 2008; Marti, 1991). Post-partum women often run differently because of biomechanical changes that occur during pregnancy, such as instability in the pelvis. Women are more prone to knee injuries when their quadriceps tire, making assessment of how much soreness is acceptable important (Burnett, Smith, Smeltzer, Young, & Burns, 2010). Almost every woman I talked with suffered from a running-related injury at some point, from minor strains and sore muscles to tendonitis to IT band syndrome to plantar fasciitis to fractures and bursitis. "The goal is to run past your body with your mind; Running is painful." Veronica said that running hurts, and she doesn't feel good until mile 6, so running is about using the mind to overcome the body. A few women I talked with thought they were or had been addicted to running in unhealthy ways as evidenced in injuries, physical and psychic. "Incitements to run harder and push through pain are in harmony with the constant exhortations to women: Regulate your weight and appearance! Ignore hunger, ignore suffering, ignore anger!" (Menzies-Pike, 2017, p. 217–218).

"Running Is Different Now," Sabrina Said

> I started when my first marriage was unhappy. I was addicted,
> and it was unhealthy.
> I loved it; I would run 6 miles 4 days a week.
> I married a high school boy, and I was young. I was happy.
> Unhappy. Happy.

Now I run because it helps me destress.

(I ran a 54-minute 10K back then. No way to do that now!)

My husband now was running a bar, so I got into and lived that life

and was 230 pounds. I decided I needed to do something-

I wasn't unhappy, but I decided to try running again 18 months ago

and have lost 60 pounds (down to 172 pounds).

I am not thin, but I am strong.

There is something about turning 40. I know many middle-aged women runners.

Like Sabrina, Kjerstin told me that running had brought her many injuries. "I have had so many injuries, and I think that's just a part of being an athlete." The list of injuries is impressive:

Injury: *Exertional compartment syndrome* in high school.

    Prognosis: Surgery or stop running.

    Action: Stopped running in college and took up water polo.

Injury: *Tore* left and right shoulder and *broke* a vertebra playing water polo.

    Prognosis: Surgery to fix shoulders.

    Action: Surgery and started spin class in graduate school;

    started running again after signing up for first triathlon.

Injury: *IT band syndrome*. Recurrence of *Exertional compartmental syndrome*.

    Prognosis: Rest.

    Action: Careful training and small increments in weekly mileage.

Injury: *Broken* metatarsal, *stress fracture* because of tripping while trail running.

    Prognosis: After not going to the MD for a year ... surgery for the

    chipped piece of bone hanging from foot.

Injuries: *Osteoarthritis, Amenorrhea, Sweet's syndrome.*

    Prognosis: I have had problems off and on with amenorrhea since I was a teenager as a result of my participation in sports. I also got a crazy autoimmune disease, Sweet's syndrome, which may be a precursor to rheumatoid arthritis. It isn't necessarily caused by running, but I think running contributed to my body's stress level and stress seems to trigger flare ups. The first time it happened it was terrible—I got this huge, painful rash, and joint inflammation. I was working night shifts, doing triathlon training 15–20 hours a week, and in a long distance relationship that meant I was away from home all of the time.

    Actions: Keep running. Right now, knock on wood, I am doing well, though. I did fall off the treadmill last month, while I was sick with an *E. Coli infection* and *broke* a rib and got *scraped* up. But I am not counting that as a real running injury—it was more from *fatigue*. I probably shouldn't have been running that day. It also didn't stop me from running at all—I just got back up and finished my run. I had some pain from the run for a bit, but it didn't limit my running ability.

I had to take it a little easy and back off on my pace for a few weeks due to fatigue and dehydration but not from the rib.

"Running Is My Therapy," Said Elyse (et al.)

Because when everything else is chaotic and out of control, my run is MINE. (Cat)

Women claimed that running saved their lives, because of the way that running helped them to lose weight, but perhaps more important were the mental aspects of running. There are both short- and long-term benefits of running for women that include coping with personal crises (Leedy, 2009). Rachel said, "People don't really talk about the emotions involved in running. Maybe they're too personal, or too complex. I work out emotions when I run. I can be almost overwhelmingly happy, and I've also cried on runs." Sarah liked running, because she could feel anything while running. The emotional release was positive, whether she cried, laughed or raged. More than one woman I talked with used these words—*Running saved my life*—without hyperbole. For example, we have Rose's story:

I am 46, live in Louisville, have run on and off for years, but more seriously since 2010. I finished my first half marathon in October 2010 and since then have run multiple halves, 10Ks, and even a full marathon. Along the way some hidden health issues cropped up, and I have had to give up gluten. I am slowly finding my way back to being healthy. My husband insists that *running saved my life*.

Running for women was a way to contend with difficult relationships, with work and relationship stress, to get solitude, to meet new people, to be outside and aware of one's surroundings, and to deal with relational loss as in Teota's case (c.f. Leedy, 2009).

Teota's Story

I had lost control and identity when my son was killed, running has given me both. My son was killed in Iraq on 3/3/07 by an IED [improvised explosive device].

B was my only child.
My life fell apart.
    I gained 60 lbs, topping out at 227 lbs.
I got up every day and went to work,
    but nothing else.

I lived on antidepressants and therapy sessions.

Three years ago, I met another woman who had lost her son in the war. She and I became pen pals:

We shared our feelings openly and began trying to understand and find a way out of the grief.

My friend suggested that we run the Marine Corps Marathon in DC as a way to honor our sons.

It was not uncommon for the two of us to come up with crazy ideas

while in the antidepressant haze, but this was the craziest.

Since I had never run, was extremely overweight and could barely get myself to work each day, the prospect of running 26 miles was nowhere in sight, but she wouldn't quit pushing the idea.

She would boast "you can't cry when you run or you'll blow snot."

I finally agreed to seek a trainer, knowing they would all tell me that dream had passed and that it would be dangerous for me to consider such a goal, but that's not what happened.

The first group I called (Jeff Galloway) had a director that insisted she could get me ready for the Marine Corps marathon. I was frustrated,

because I really wanted them to tell me no!

But it also gave me a little hope.

She and I finished that marathon and there was something euphoric that flooded over me when I crossed that finish line and the 30 other marathon finish lines since.

Running as Accountability

I was once a very unhealthy person and a heavy smoker. I've become a different version of myself because of running. I never want to be who I was, and running keeps me accountable to myself. (Rachel)

Some people, when they hear about my running habit, say, "I could never do a marathon." Hooey, I say. Anyone can run a marathon. It requires a commitment of time, energy, and a willingness to get through the parts that aren't fun. You need to run when you don't feel like it. You need to set a reasonable, reachable goal. You need to break it down into manageable chunks and build up your strength. (Toor, 2014, para. 13)

One of the biggest benefits of running and what women considered the best thing about running was the way in which running kept them accountable to themselves and their goals: weight loss, weight maintenance, self-care, physical health, mental health, running faster and further, running better, training for races, seeing what you can do with your body, staying connected, staying strong and independent, being a better relational partner, and being a better mother. "I want to be 82 years old in a running skirt running a race" (Patti). Jaclyn told me:

> With running, you are in competition with yourself. If I get too wrapped up in watching what the people in front of me, beside me, or behind me are doing, it throws off my entire mentality. It is just me looking at myself from race to race and seeing what I can improve, thinking about getting a PR, or going out stronger, or running longer without a walk break. It has showed me how to be competitive with myself, but it has also showed me how to be positive. And that even though I did a training run that didn't go how I wanted it to, if I finish it then that, in and of itself, is a victory.

Some women used online means to be accountable and stay motivated (see Chapter 4 for a discussion of women running online). This could be logging miles in Facebook groups devoted to running, using a Garmin, posting runs on social media, and using running apps that keep track of mileage (e.g., Daily Mile, RunKeeper). Runners told me that they liked how using apps and social media let them encourage other runners, too. Some ran on treadmills to keep track of their running data and ran with friends and running groups. Of course, these positives were not so for everyone. Jess said running was not effortless, was too solo, and that is why she does not like it. "You have to be disciplined," she said. She once ran with a training group for a race she wasn't running to get in cardiovascular work for biking, which is her main sport. While she liked the group aspect of the training group, she disliked when anyone who ran over a 12-minute mile was considered a "walker" and placed in that grouping. She questioned definitions of "runners" as those who ran under 10-minute miles. And while she disliked the idea of competition, the effort, and the discipline required, many other women told me that was exactly what they liked about running.

Is this what you like most about running? The exploration and setting of goals? What else? That is, what keeps you running?

> I first started running to help myself lose weight. Then I found out I really enjoyed the running itself. I think the thing that makes running so meaningful to me is that I enjoy it on so many different levels, and in so many different ways. Adventure is one, but not all runs are adventurous.
> I think the thing that really keeps me going, day after day—the thing that really motivates me—is the way running gives structure to my life.
> I love the goal-setting, though. That is probably a close second.
> Then the adventure and social aspects are the icing on the cake.
> I'm a mess if I quit running for more than a couple days at a time. (Marsha)

Running as Relational Practice

> Every time I go out for a run, I'm modeling dedication and a healthy lifestyle choice for my kids. (Rachel)

Women's running practice was relational. Some considered their running as strong modeling of healthy behavior for friends, children, and others in their social networks. Most of the women with children explained how their running was good for them and their child:

Hi Sandra,

I was thinking of your work today as I ran the charity 5K I told you about during our interview. My 8-year-old daughter ran the 1-mile kids' run and placed first! She was so proud of herself for putting her heart and soul into it. It really got me thinking about how much of my running has changed to being a role model for her about healthy women's bodies and having a good sense of self and self-confidence. She was so proud of herself, and it was wonderful to see it related to confidence and athleticism instilled in her at a young age, hopefully, from seeing her mom as a role model for a healthy lifestyle and state of mind rather than seeing a mom worrying about weight or dieting, which mothers and women often get stereotypically associated with. I'm not sure I included that in the interview, but I think it reflects an important change in the way I view running in my 30s as compared to my 20s!
I hope you are doing well and all of your other interviews are going well.
Best,
April

Women received a generous amount of support, mediocre support, or no support from significant others in their lives for their running. Sarah always knew where her fiancé was in a crowd; he would use the duck call to cheer her across the finish. In general, some research suggests families use egalitarian and cooperative strategies to deal with conflict about time spent running (Goodsell & Harris, 2011). Training for marathons is difficult because of the time commitment; women often have to leave their families to train, unless a family member is running with them. And as Alex told me, others mention women runners' time away from family more than men's time away. It is never "men are spending time away from their family to run."

Q. How do significant others support (or not support) your running?

A: No one else in my family is an athlete and my mother in particular gets upset about me running—particularly when I am injured or having problems with amenorrhea. I feel like she often begrudgingly supports me or tolerates me running most of the time, but I try to not talk about it and do not expect or ask for any support from my family or really anyone else in my life. (Kjerstin)

A: My husband supports my running in a sort of passive way. Going out to run is just something I do, but he's never come to a race, except my first. I think he's as supportive as a non-runner can be. He doesn't understand the accomplishment I feel. My mother thinks that I should be skinny because I run, even though I am not overweight. She comments on my body in negative ways and dismisses my running as anything significant or worthwhile. (Rachel)

Q: What role (if any) do others play in your running?

A: Other people don't play a role in my running. As a mom, most of the choices I make throughout any given day are for my family, so running is a choice I make for myself every time I tie on my shoes. (Rachel)

Research supports women's experiences. Women who run long distances gain social benefits like a sense of community and belonging and an enhancement of self-esteem (Shipway & Holloway, 2010). Novice women runners who train for long distances often experience marked positive changes in their relationships with self and others such as increased feelings of empowerment, openness, determination, ability to cope with stress and depression, and achieving a beneficial mental outlook (Boudreau & Giorgi, 2010; Leedy, 2009).

Running with and without Others

Running can enhance personal relationships (Boudreau & Giorgi, 2010). Elyse told me how running with her father, who was a high-school track coach, was a form of bonding. They got to talk about important issues during runs: "He is such a good role model, and he tried to talk about all kinds of things with me. This goes against the stereotype of men raising daughters." Jaclyn got to know her neighbor better when they began to run together. Others told me how they ran with friends, their husbands and wives, their dogs, running groups, and their children. Pamela ran with a chapter of *Frontrunners*, an international LGBT running and walking group.

Of course, running can be a challenge in relationships. Terra thought that most women would not want to run if someone in a running group runs faster or is too competitive, because you have to stay in the pack. Elyse also told me that she had a romantic relationship end when her male partner wouldn't run with her because she ran faster than he did. And he believed that men should run faster than women. Jess felt that her mother used running as a means of ignoring problems in her life, and she resented the amount of time her mother, who came to running in her late 40s, spent "running away from her problems."

Q: *Do you run with others?*

A: Almost never.

A: I run solo. One of my most favorite things, though, is to run long solo runs in the mountains. And I love the feeling of moving across the landscape. I like to set a goal that at first seems unmanageable—something I can barely get my head around. I have run some trans-Sierra routes that were like that for me. I start out a little scared and slowly let the run unfold. (Marsha)

A: I do run with others occasionally. I have been a part of several triathlon training groups who I ran and trained with. I also run with friends and coworkers. Generally, I have specific workouts to do and I wouldn't compromise them to run with someone, but if I can find someone who can do them with me I like the company. I also love using my easy runs to connect with my friends and encourage them in their running. (Kjerstin)

A: I mostly run alone. Or I take my dog if it is a short run. I am part of a group on Facebook called Rubenesque Runners. It is for women who love art, food, and running. The ladies in the group post frequently about running books, running topics, and when/where they will be running if they are looking for partners to join them. I have joined them a few times for some longer training runs. It is really nice to have someone to chat with along the way. It passes the time much quicker. I don't like being so dependent on someone else for my pace, however. I like stopping when I want to stop and running at the pace I need. I haven't found a running partner who matches my pace yet. (Jessica)

Running as Safety and Danger

Umpteen variants of this story operate around the core of a woman being punished—through violence, mockery, exile—for running in the wrong place, alone and without protection. It's never neutral for women to run. (Menzies-Pike, 2017, p. 128)

Q: *What does it mean to be a woman who runs?*

A: Being able to run in parks without fear of molestation, whether that's to train for a marathon or to get a bit fitter, is part of a bigger freedom to be safe in both public and private places. The decision to run 5 or 10 or 20 miles is the recognition that our bodies are our own, and that we can choose how far we run, whom we sleep with, what we eat, whether or not to take a pregnancy to term, and how we might swing our arms and legs to take us through our days. (Menzies-Pike, 2017, p. 21)

Being safe as a woman runner means being aware of potential danger in your surroundings, running in groups, running inside, not running too fast or too hard in case you need to defend yourself, remembering you have a "big mouth and not

74 Real Women Run

telling dudes to stop staring at your chest because you don't want to get into a fight," running with pepper spray or a gun, not running with ear-buds, running with a dog: BEING HYPER-VIGILANT. Kristin ran in Chicago and had been stalked twice before, so running for her was about being safe.

> Everything is proactive for me. Safety is a big issue for me; I don't want to become a statistic at all. I want to combat most of the violence—it's not just the physical violence but the verbal violence. You know people yelling at you, but I'm prepared to fight. I try to be mindful of that.

Sarah disliked having to think about safety. She carries pepper spray on many of her runs, and does not run at night. "Women do not treat men that way." She knows that if she does not wear a shirt over her bra while running that she will be ogled. "And I sweat a lot. One time, a man slowed down with his car window down, and tried to talk to me. I just kept running, and I had my finger on the pepper spray."

Catcalls

> I usually run at a park with a 3.3-mile loop. I was finishing up the loop yesterday, feeling strong (coming back from surgery 6 weeks ago), and I was thinking about how happy I was that I went out for that run, since I'd just run the same route the day before. Immediately after having that happy thought, a man hanging out with buddies in the woods whistled at me, throwing me out of my zone. I was incredibly upset, angry that I couldn't just be left alone in a public place, angry that I was on display, apparently. And, angry that there was nothing I could do. Confrontation could be dangerous. (Rachel)

A few women told me that catcalls were their least favorite thing about running. "You can run, but you can't escape sexism." (Kita, 2016, para. 6). "Getting yelled at and catcalled by drunk frat boys on their porches is one of my least favorite elements of running outside." Emily told me she finds catcalls irritating and distracting. She runs in the city and sometimes near a college campus. Emily sorted out catcalls from supportive calls from bystanders like "Way to go. Nice job." She tries not to react to the catcalls and keeps running. "If they say something rude, I will shoot them a dirty look." I was pleased to see an article in *Runner's World* wherein Meghan Kita (2016) tells a story about sitting in a planning meeting for the *Zelle*, a female-focused channel at *Runner's World*, and talking with fellow women about catcalls and harassment while running. The shock of her male colleagues got her to challenge the idea that the "problem" is women running alone, rather than sexism and generic safety tips. The highlighting of women attacked and killed while running exacerbates the issue and plays into fear and ideas that women are responsible for harassment and violence.

> Men, I want you to think about this. I want you to care about this. I want you to be outraged that the women you run with, when they're running alone or even with other women, are angry but not surprised when men whistle or honk or make lewd comments. Awareness and understanding from our male counterparts is more likely to improve the reality of being a woman who runs than a whole browser full of hackneyed "safety tips" ever will. (Kita, 2016, para. 17)

Running for a Reason: Running for Remembrance

> Sometimes I run for my dad, who has ALS and is about to get a wheelchair, because he's losing his ability to walk. (Rachel)

Women felt that running communities were supportive. "Running is a sport where all are welcome" (Jaclyn). And it is this sense of community that led many women to run races for causes. Women run to raise money for charity, to raise awareness of social issues and diseases with running groups such as *Back on My Feet* (www. backonmyfeet.org), which is a running mentor program to combat homelessness. Some run to remember loved ones killed in war, like the group *Wear Blue*, which is

> a national nonprofit running community that honors the service and sacrifice of the American military. Wear Blue: Run to Remember creates a support network for military members and their families; it bridges the gap between military and civilian communities and it creates a living memorial for our country's fallen military members. (www.wearblueruntoremember.org/)

Alex runs 100 miles a month for *Wear Blue* in her position as a nurse on an army base. Danielle also runs 100 miles a month in the Century Club as part of *Wear Blue*. She also uses the Charity Miles app, in which one presses start when going out for a run to raise money for a charity. Danielle runs for the *Wounded Warrior Project*, earning 25–50 cents a mile. Others ran for *Team Red, White, and Blue*, whose "mission is to enrich the lives of America's veterans by connecting them to their community through physical and social activity" (www.teamrwb. org/). Jaclyn liked to run in races that were for local causes like the animal shelter in the town where she lives.

Back on My Feet

> About 22 weeks ago (that's 670 miles ago), I started to train for my first ultra marathon. I also joined Back on My Feet (BOMF) that same week the

beginning of June. My life has been changed since then. I have met some wonderful new friends and running buddies. I am inspired by my friend who battled and beat alcoholism and who ran for the first time in his life. His first run with BOMF we ran together, and last month he ran his first half marathon. He is only one of the several amazing people that I have met with BOMF that have overcame challenges and set-backs, which set themselves up for success (you know who you are—love to you)! I've met and become close with running volunteers that share similar passions. I've met a volunteer who shared his story with me about how running saved his life after his son passed away from cancer. I've met a few friends who were feeling discouraged with running and BOMF rekindled the flames (congrats C. who slaughtered his first half marathon last week and shout out to J. who lives right across the street from me—and that makes meeting up for runs so easy and fun). The stories can go on and on. I've never been a part of a running group, and I am so happy to be part of this one that focuses on running and makes positive change. Always need to thank K. who introduced me to this group. Good luck with the NY marathon! Love to all of you!

I am also so happy to have met my long-distance runner friend A. randomly on the lakefront path a few weeks back. I look forward to seeing you on the race trail. I am so proud of you doing the almost impossible of 50 miles. For those of you that I have run with over the summer or the past several years, you all continue to inspire me and be part of my running story. And you all will forever be in my heart.

I think I get emotional (in a good way) before a race. Maybe it's like a reverse PMS for runners. It seems like it's a time to reflect and give pause before I start.

Yes, I am excited. Yes, I am anxious. Yes, I am a little intense before races! Yes, I feel ready. I feel so blessed to have such beautiful friends and support in my life. Thank you all for your encouragement, inspiration and embracing my running craziness! You will be in my heart while I run mile after mile after mile on Saturday. (Shanda, Facebook Post)

Running in Races

I have to have a race to work towards, or I won't run. Races are my favorite part. I'm not that fast. Slower than average. I hate paying registration fees to run in a race, but there's just something about pinning that bib on and hearing the gun. I love meeting new people. I love passing people. I don't like it when others pass me. I love chatting with a fellow runner for a few miles and then moving on to someone else. I love hearing their stories … encouraging a newbie on the road … seeing foolish people who are running too fast past me and then passing them later. (Jessica)

Women I talked with defined runners as those who race, and specifically, those who run marathons. The definitions of *who* counts as a runner may be due to the number of races one competes in, to how others view one, to the process of identification. The discourse in popular culture is that real runners run, it takes a lot to be a "runner-runner," and "serious runners" all run marathons. Melissa is part of a group called "Half-Maniacs" and wants to run 50 half marathons before she is 50. At the time of our interview (5/30/2014), Melissa was 46 and had run 31 half marathons. She identifies as a runner and uses lots of gadgets (e.g., Garmin). Other women, like Kristin, aspire to be a "runner" or talk about themselves as average or slow runners. Kristin runs to stay healthy and be fit for a pregnancy she wants in the future. "Perhaps someday I will run a race, but for now, my goal is to run 1 kilometer a day." Patti told the story of how her first race was a marathon. She googled how to run a marathon and found a man who ate snickers bites as fuel. When she ran her marathon, she packed four full-size snicker's bars and ate one every 6 miles. She laughed at her mistake and said it was the worst thing ever to fuel her run.

Women told me they raced to have something to work toward, to keep themselves training and running, to be social with other runners, for the camaraderie, to feel like a runner, to get race shirts, medals, bags, and other race premiums and bling, to challenge themselves, to run for a cause, and to meet their goals. And some runners, like Sarah, won races in their age groups. Others like Patti, were happy to be "back of the packers."

> In endurance running … one thinks at great length while doing the activity. To run 5 or 10 or 26 miles is, as much as anything else, to engage in a sustained way with the deep strangeness that is the human mind. (Schulz, 2015, para. 2)

Running in races is a way to stay motivated and expand their running selves.

> I find it highly motivating to race, so I am almost always training for SOMETHING. I just raced a 50k on Saturday, and next is the 500k Vol-State. I love the social aspect of races. Some I just do at a leisurely pace for the fun of it. Other times, I really enjoy the challenge of pushing myself, training hard, and focusing on the challenge at hand. (Marsha)

Races are also a way for women to connect with other runners, to feel part of a community, and maybe talk to someone you never would have otherwise. Jaclyn recalled a race she ran for a charity involved with stroke victims, where she met a stroke survivor who was running. "Races are good because you get to talk to people that you usually wouldn't and hear their stories." There are running groups on and offline that support, encourage, and provide community, such as Half-Maniacs, Wear Blue, and Marathon Maniacs. Some women were part of a Century

Club, where the goal is to run 100 miles a month to total 1,000 miles a year. Other women told me they do not race because of the money, time, and stress.

> Q: What do you like about races?
>
> A: I love the excitement and the sense of community. (And the bling!) I like being around people who like doing what I like doing. Runners are typically friendly and good people. (Rachel)
>
> A: I definitely race! I generally don't enjoy the actual race, but I love the focus and purpose that it brings to my training. Finishing and setting new PRs [personal records] makes all the pain and suffering of racing worth it, too. The camaraderie of racing is also really neat. I like meeting other runners and watching them run. I absolutely love having friends race with me, too. I don't really race them individually—I am always just racing myself, but it's really fun to support each other. (Kjerstin)
>
> A: I do race! I like racing because it tells me how I am doing with my training. I like reaching new PRs, and I like the purpose it brings to my training. It makes running more fun :) (Tammi)

Wednesday, June 11, 2014 at 5:55 PM   *Re: Runner Study*
Hi Sandra,
    I remembered something that might be of interest. When you asked me why I participate in races, a complete answer hadn't occurred to me. Not surprisingly, I don't expect to win when racing, but I do expect some sort of validation for all of my running "work." Basically, doing well in a race (meeting some kind of goal) proves to me that I am strong enough, maybe even stronger than I expected.
    So, while I do like to pass other runners and place higher, what I like even more is surviving the hardship of a race better than I expected. Pushing myself to deal with whatever challenge comes up during the hours of a race is perhaps the biggest incentive to race.
    That's all! It was fun talking to you.
Veronica

Running with Your Body

> For feminists, questions about the mind and body aren't just pathways to running faster—they're also a way to understand the hierarchies that have structured gender. Controlling bodies is exactly what patriarchy does to women and queer people. (Menzies-Pike, 2017, p. 217–218)

I asked women about their running bodies, and they showed me pictures of themselves sweaty and accomplished with race medals crowning their necks, pictures of them with running buddies and family, on a favorite running path, at the beach, in the woods, somewhere around the globe, before, during and after a run, at the finish line. They told me these pictures showed how strong they were, a moment when they felt good, felt part of a community, and felt proud being a runner.

Flying Pig Half Marathon, May 4, 2015

For someone who lives in the mind, running reminds me and forces me to be in my body, to pay attention to my movement in the world, to know that my steps matter, to focus on having to pee and the twinge in my arches and to get to the space where I am a runner.

And a runner is all air and urine and dodging and you are running, so you cannot eat for an hour after a run because you eat air and music and yelling.

> The joy of movement
> being in my body this moment
> is a radical workshop. (Field Notes)

Women liked how running changed their bodies and how they felt in them. "You are in your body when running in ways you aren't the rest of your day. When running you are aware of your breathing, your feet on the pavement, and your body" (Emily). Kristin said that running changed her metabolism and allowed her to eat cheese. Danielle commented that having strong powerful legs—Thunder Thighs—was a good thing. Sarah felt running reminded her to "listen to your body. Love your Body. Enjoy the ride." She had anorexia at 18, and running allowed her to like her body and to focus on what it could do. For the first time in 10 years, running let her enjoy her life. Elyse also talked about how running bodies were not just thin, slim-hipped, and small-breasted women; women who run have strong legs and slightly bloated midriffs. She was experiencing tearing in her breast tissue from running, but felt dissatisfied with the gynecologist's response to quit running or have breast reduction surgery as neither of those options was appealing. She wore two sports bras to run in, and that seemed to help. Pam told me she had breast reduction surgery, and that made running and exercising much better for her. Em described how running for her was a way to attract a partner and keep the body she wanted:

> I run until I am deliciously sore. I run to be strong and to be in the body I want. I am single, and I may be for the rest of my life. When I was running in Africa in the Peace Corps it was the best I have ever felt. I got a boyfriend, because I was strong … I get frustrated with my mom and others who think you need a man to do things. Once, she wanted to call over a neighbor to help move some seats out of a mini-van. I did it myself. I want to be able to do things without a man, and running makes me strong.

Q. How would you describe your running body? What are your strengths? What do you like about your running body?

A: I love my running body. Strong legs. Sculpted calf muscles. The split muscles on my thighs just above my knee. My strengths? Not my stomach (thank you three little boys). I have a nice relaxed stride. When I am training for a long distance, I feel slim and toned and AMAZING. (Jessica)

Q: How would you describe your running body?

A: My running body is more of a regular person body with really strong legs and some giggly mommy areas. (Rachel)

Q: How would you describe your running body?

A: I am not long and lean. I am not fit and trim like the women in the running magazines. But it's [my body] running. (Pam)

Run Body

I like how running gives you a different perspective on what your body is:

Every day I work on being stronger and faster
    not on being skinnier or prettier.
I could be the ugliest girl out there, and if I can run well,
    I would still be happy because of the way it makes me feel to run.
I feel strong, powerful and beautiful when I run
    even though any race photo would show me that I actually look
 like a hot mess!
I try to look at my running body as a machine.
    I give it the fuel and care that it needs so that it is ready to go every
morning when I wake up to run." (Kjerstin)

All women mentioned their bodies in relation to what they and others considered to be "running bodies." That is, a running body was akin to a narrow ideal of elite athletes: slim hipped, athletically thin, toned, no evidence of having birthed, no stretch marks, no curves or jiggly loose flesh. Runners "glide and take long strides." Women went on to argue that running is for everyone, though. They described the variety of shapes and sizes they saw at races, and in their local running spaces. "You can be at a race and see all kinds of bodies. You will line up with 80-year-olds" (Jaclyn). Pam liked the variety of women's bodies she saw at the back of the pack during the Marine Corps Marathons.

Women were also aware of how others evaluated their bodies. "Going for a run shouldn't feel like a performance. Often, it does. Women get looked at—ogled, appraised, admired, gazed upon with adoration, spied on—all day and all night" (Menzies-Pike, 2017, p. 107). For example, Kjerstin told me:

> There is an older man who bugs me while I am running at the gym. The other day he told me I had better not run too much because I was a pretty girl, but if I wasn't careful I would start to look like a boy.

Jessica avoided talking about her body because "my runner friends all have that ideal runner's build—tall, lean and leggy. I'm not. I'm overweight and pear shaped. I might feel intimidated by them if I talked about these things."

Jessica's Story: Never Good Enough

> During my weight loss/running journey, I lost 78 pounds.
> I dropped from a size 18 to a size 6.
> When I went to both of my marathons, I felt overweight! When I run in a shorter race, there are lots of other people who look like me …
> not exactly like a distance runner. But when I ran in my marathons, there were no overweight people around me.
> I was a size 6 and should have been happy and felt great. I had worked so hard to get my new body.
> But I was surrounded by those naturally slender people who can eat ANYTHING they want and still look like supermodels.
> No one else looked like they carried any evidence of having carried/birthed children or breastfed babies or had leaky bladders.
> It was a bit depressing; I have to admit. And I will never look like those people, no matter how much weight I lose or how much running I do.
> I wasn't given that body type.

Good and Bad Runs

I asked women how they feel when they run, what a good run feels like, what a bad run feels like. The determination often depended on the effort required, and whether women met their goals for that run. For most women, running required a lot of effort. Some women told me they never felt it was easy.

> Somedays it feels amazing and some days everything hurts and my legs are cranky and it feels awful. Sometimes it's a mix of both :)
> Some days I wake up excited to run, other days I have to kick myself out of bed but as soon as I start; it's rare that I am not happy to be running. I am always happy after a run is done although I do occasionally put in extra mileage because I can't stop!

82   Real Women Run

Jaclyn told me that running is a balance of good and bad. "You are going to have bad days. You just need the mentality to push through that. And look forward to that one good run. When you train well, you know when to take time off."

Jessica Takes a Run

Running is a menagerie of feelings and emotions.

Before a run, I sometimes feel dread or avoidance. Sometimes I can't sleep at night, because I'm so nervous/excited/anxious for a 20-mile training run early the next morning. I get shaky in anticipation of the run at times.

I love standing on my back step, stretching my calf muscles, turning on my music, starting my stop watch and taking those first few steps.

During the first few blocks, I feel awkward. My shorts aren't in the right spot. My headphone wires are not tucked in properly. My water belt is not tight enough. My shoelace is too tight on one foot and sticking out on the side of the other foot. I worry that I did not put enough Body Glide on my "prone to blister" spot on the inside of my thigh. Then I find my stride, I start to glide over the ground.

The first mile feels okay.

Mile two is okay.

I hate mile 3. Every time. Despise it. I can't understand why I keep doing this to myself.

My muscles are tight, my breathing is not right, and I just want to go back to bed. Then, I get past the hurdle ... that horrible third mile.

Things start to get easier. Miles 4, 5, 6 improve.

My favorite mile arrives—mile 7. I guess this is my runner's high.

I love mile 7.

I've even gone past my training mileage a few times in order to get to mile 7 and feel that feeling.

Nothing can stop me. I look great. I feel great.

Other people driving down the road must be admiring me. I no longer care that my abs might be flopping over my waistband with each step. If it's a long run, the adrenaline keeps me high ... and when I know I have just a mile left, I try to really push hard.

Pick up the pace. Go past that burn in my legs that has something to do with lactic acid.

When I hit that last corner and can see my mailbox, I fly like the wind.

I'm probably not moving very fast with whatever is left in my tank, but I feel like I'm sprinting like the best.

After I feel great. I keep walking, usually a block or so to cool down. I am exhausted. I'm thirsty. My bladder is full. My feet are yearning to be free of the socks that confine them. I am hungry.

I conquered the world.

I followed my training plan. I accomplished my goal. I am one step closer to that marathon. I can't wait to strip off my clothes and dive in the swimming pool and drink my smoothie.

Good Run   Bad Run

Can You Describe a Good Run?

A good run is when I feel strong and light. Usually after 4 miles I hit a sweet spot for at least 2 more miles where I'm faster, my breath is easy, and I feel like I'm flying! That feeling is what I'm chasing now. (Jessica)

A good run could be good because it was fun, because I felt good, or because I felt terrible but I still stuck it out and did my work out how I was supposed to. There are a lot of reasons that I would enjoy or be happy with a run. (Kjerstin)

I like to run until I can't move and feel like my legs will fall off like when I was a kid. I want to be comfortable in my body and get back to the evolution feeling—a baby crawling, walking, then running. (Kristin)

A good run is when I sweat from effort, when I smell. (Sarah)

Can You Describe a Bad Run?

A bad run is a run after a night of heavy food or beer. It's a heavy, slow run, with aches and pains, and heavy breathing at a slow pace. Or, a bad run is that little voice inside that says I'm too tired or I can't do it.

Is there such a thing?

A bad run would be one where I gave into fatigue or pain and ran slower than I wanted to or not as far although I very rarely allow myself to shorten a run unless I have a legitimate illness or injury.

I tripped on the side of the road and twisted my ankle on a 15-mile training run. I got up, it hurt like crazy, and I had 3 miles to go. My husband and my three sons were picking me up at the end of my run. I had no phone. I just ran through the pain. By the time I arrive and looked at my children's astonished faces, I looked down to see that I was bleeding and covered with road dirt from temple to ankle. I have road 'burn' on the right side of my face, my right shoulder and arm, and all down the side of my right leg. And my ankle was the size of a very large cantaloupe. I probably shouldn't have

84   Real Women Run

run on it, but I didn't know how bad it was! I had a swollen ankle for more than 2 weeks and had to take several days off before my marathon. It was very hard to see those days go by and miss those runs. And then I gradually got back into it and everything was fine in the end. (Jessica)

A bad run is when I walk and do not meet a PR or the goal I have, when I think about how bad I feel. (Jaclyn)

A bad run is days when things are not going smoothly. When I literally run my shoes into the ground and having to buy new ones. (Jaclyn)

A bad run is when I can't release my stress, and I am worried about a training plan. Like I need to take a 30-minute run, and I have 29 minutes. (Alex)

Running as Expansion

It is after a good run that I feel I can give something back to other people. I have more patience, more love to give. (Marsha)

Women ran to change their minds, to feel strong, to connect body and mind, to forget mind and focus on body, to build relationships, to run away from bad relationships and versions of themselves and to run toward self-definition. "Running is like fighting against myself and winning" (Mavis). Running shaped who women are and who they want to be. Pam said that "People can do more than they think they can do." That is why she liked running. Kristin described running as a process like being born: "I think better and am more awake after I run," Jackie said "Running is about racing against the self. You can take yourself out with negativity or you can build yourself up with positivity. When others ask me, should I try running? I tell them it is worth one event." Terra found running changed her brain, made her do things she would never do as an introvert, like talk to others and jump in a race and encourage others. "You can do hard things." That is what Shireen likes about running. I listened to her interview as I ran 8 miles with my jacked-up shoulder, knee pain, and an achy toe. She said that what she likes about running is the difficulty, because it reminds her she can do hard things. And a good run is rare, one that is not complicated, where you are not thinking about your body, you are talking with friends, and before you know it, you ran 8 miles.

Most runners run to achieve either or both of these conditions as often as possible—to provoke a kind of Cartesian collapse, mind and body suddenly in anguished or glorious collusion. And, most of the time, we fail. The body twinges and hitches and aches; the mind fusses and fidgets. What is it all for? What was it all about? (Schulz, 2015, para. 18, 19)

Running Toward ... Running Away

The other day an acquaintance who knows how much I run asked me: what am I running from? As I ran that evening, I realized that I don't run to run from. Quite opposite: I run to.

Run to gain balance, perspective, release, relief, freedom, physicality, spirituality. Run to see, run to be happy, run to focus, run to challenge, run to expand my mind, run to explore limits, run to give back, run to create bonds/friendships, run to inspire, run to be inspired, run to live, run to be me. (Shanda, Facebook Post)

Q: What does running mean to you?

A: It's the glue that holds my life together.(Marsha)

Q: What does running mean to you?

A: It is an outward sign of discipline. An inward sign of my personal journey towards wellness. It symbolizes life—peaks and valleys, low moments and high moments. It should keep me healthy in the future. It should inspire my children to do something. (Jessica)

Q: What does running mean to you? For your daily life? For your future?

A: Running means so much to me—it's hard to even put it into words. My day just isn't right if it didn't start with a run. It's my time to think, to dream, to relax, to be happy. My social life revolves around it, it influences how I spend my time, organize my day, what I wear, what I eat, how I feel, what time I go to bed—so much of my life is influenced by running.

As far as my future goes—I hope I will be that 90-year-old lady kicking the 20-year-old girls' butts out on the race course!! *I hope I can run forever and that it will continue to be a part of my life.* I also know that may not be possible and try to look at every day I can run as a gift, because I don't know when I will no longer be able to run. It could be tomorrow, or I could run until the day I die. (Kjerstin)

Running as Self-Definition

It seems that through running, you are actually relearning everything that you came into the world with. (Andrea England, 2017, personal communication)

86 Real Women Run

A runner is …

A runner is one who enjoys running, and sometimes hates running. A runner is one who walks and jogs and sprints and runs races and does not run races. Women exhorted the right to self-define. And once women thought about what running meant to them and their embodied selves, then the importance of labeling themselves as runners, regardless of their ability, interest, and pace was apparent.

"I am a wife and a mother, that is what I do. Being a runner is who I am." (Mavis)

> I like finishing strong. I like hitting mile 7. I like knowing that I can do something that other people only dream about. I like working towards a goal—a race. I do not like taking the time to actually run. I do not like that I lose my conditioning fast if I neglect my running. I do not like running the same route over and over again. I do not like the ordinary running schedule—a few miles each day. It is monotonous. I do not like running if I do not have a big goal in mind. This might be why I don't consider myself a runner all the time. (Jessica)

> Q: What makes one a runner?

> A: Running. (Rachel)

Woman, Runner
a woman runner
is a woman who runs is
a woman runner

> Q: What makes one a runner?

> A: I was a runner. Or maybe not. I am not currently running. I don't know if I love to run. Sometimes it is a chore. Sometimes it is exciting. I have a love/hate relationship with running. Right now I hate it … and I am lazy. A runner is … someone who runs consistently. Who loves it, craves it, and wants to do it. (Jessica)

> Q: What makes one a runner?

> A: If you run you are a runner. I think it's important that you see yourself as a runner, whether you just run a few miles a week or if you are a pro! People often tell me they really aren't runners, but in my opinion you are cheating yourself if you don't own your identity as an athlete. (Kjerstin)

> Some athletes like to talk about what a simple sport running us. They say that all you need is a pair of sneakers. That's not true. What you need is some

freedom of movement and the ability to see a clear path ahead of you. It took me years to see that path and to find my pace. (Menzies-Pike, 2017, p. 4)

Q: What does running mean to you?

Running means

Solo Time	Me Time	Individual
Being with Others	Being Close to My Dad	Training with My Wife
Community	I Like the Communal Part	Building and Keeping Relationships
Finding a Partner	Being Social	Good Friends

Runners Are Kind.

Running means

Being Aware of Your Surroundings   Exploring   Freedom   Escape
Being Ageless   Feeling Strong   Being Grounded
Seeing New Places   Running Anywhere
Running is Mine.

Running means

Remembering   Doing for Others   Being a Role Model
Running for Meg   for Soldiers   for Mothers
Running for Those Who Can't.

Running means

My Identity   Part of Life   Empowerment   Being Determined
I'm Afraid to Stop   I've Run All of My Adult Life   I Feel like Myself
Running Is the Puzzle Piece of My Person.

Running means

Changing My Mind   Feeling Strong Mentally   Releasing Stress
Building Goals   Building Self-Esteem   Being Determined
Meeting Challenges   Weight Control
A Habit   Being Addicted   You Always Have Races
The Century Club = 100 Miles a Month
I Can Be Someone Better   I Can Depend on Running
Running Changed My Life.

88 Real Women Run

Running means

Getting Away from Bad News   An Inner Struggle   Zoning Out
Dulling Brain Chatter   Pure Physical
Freedom like Flying   Body Will Catch up with Mind
Hating Running   Liking it When It is Done   Feeling Bad when Don't Run
Running = Powerful and Strong and Confident.

Running means

Being a Better Person   The Glue That Holds Life Together
If I Can Run, What Else Can I Do?   Limitless
You Didn't Think You Could Do This, So You Are Limitless
Running is Self-Care and Self-Less.

Running means

You Can Do Difficult Things. (41 Runners)

I am a wife and a mother, that is what I do. Being a runner is who I am. (Mavis)

References

Boudreau, A. L., & Giorgi, B. (2010). The experience of self-discovery and mental change in female novice athletes in connection to marathon running. *Journal of Phenomenological Psychology, 41*(2), 234–267.

Buist, I., Bredeweg, S. W., Bessem, B., van Mechelen, W., Lemmink, K. A., & Diercks, R. L. (2008). Incidence and risk factors of running-related injuries during preparation for a 4-mile recreational running event. *British Journal of Sports Medicine, 44*, 598–604. doi: 10.1136/bjsm.2007.044677

Burnett, D., Smith, K., Smeltzer, C., Young, K., & Burns, S. (2010). Perceived muscle soreness in recreational female runners. *International Journal of Exercise Science, 3*(3), 108–116.

Chakravarty, E. F., Hubert, H. B., Lingala, V. B., & Fries, J. F. (2008). Reduced disability and mortality among aging runners: A 21-year longitudinal study. *Archives of Internal Medicine, 168*(15), 1638–1646. doi: 10.1001/archinte.168.15.1638

Dufek, J. S., Mercer, J. A., Teramoto, K., Mangus, B. C., & Freedman, J. A. (2008). Impact attenuation and variability during running in females: A lifespan investigation. *Journal of Sport Rehabilitation, 17*(3), 230–242.

Goodsell, T. L., & Harris, B. D. (2011). Family life and marathon running: Constraint, cooperation, and gender in a leisure activity. *Journal of Leisure Research, 43*(1), 80–109.

Kita, M. (2016, August 11). The problem is not women running alone. *Runner's World*. Retrieved from www.runnersworld.com/other-voices/the-problem-is-not-women-running-alone

Lee, D., Brellenthin, A. G., Thompson, P. D., Sui, X., & Lavie, C. J. (2017). Running as a key lifestyle medicine for longevity. *Progress in Cardiovascular Diseases.* doi: http://dx.doi.org/10.1016/j.pcad.2017.03.005

Lee, D., Pate, R. R., Lavie, C. J., Sui, X. Church, T. S., & Blair, S. N. (2014). Leisure-time running reduces all-cause and cardiovascular mortality risk. *Journal of the American College of Cardiology, 64*(5), 472–481.

Leedy, G. (2009). "I can't cry and run at the same time": Women's use of distance running. *Affilia: Journal of Women & Social Work, 24*(1), 80–93.

Marti, B. (1991). Health effects of recreational running in women: Some epidemiological and preventive aspects. *Sports Medicine, 11*(1), 20–51.

Menzies-Pike, C. (2017). *The long run: A memoir of loss and life in motion.* New York, NY: Crown.

Schulz, K. (2015, November 3). What we think about when we run. *The New Yorker.* Retreived from www.newyorker.com/news/sporting-scene/what-we-think-about-when-we-run

Shipway, R., & Holloway, I. (2010). Running free: Embracing a healthy lifestyle through distance running. *Perspectives in Public Health, 130*(6), 270–276.

Szabo, A., & Ábrahám, J. (2013). The psychological benefits of recreational running: A field study. *Psychology, Health & Medicine, 18*(3), 251–261. doi: D10.1080/13548506.2012.701755

Toor, R. (2014, June 23). What writing and running have in common. *The Chronicle of Higher Education.* Retrieved from http://chronicle.com/article/What-WritingRunning-Have/147193/

4

WOMEN RUNNING ONLINE

Listen to your body. Love your body. Enjoy the ride. (Sarah)

Gendered Expectations in Popular Women's Running Websites and Blogs

That's the empowered part of running—when I run, I am myself and I control myself. Nothing more, nothing less. When I put on my running clothes, I take off everything else; I leave my obligations inside, leave behind the conversations and arguments and worries of the day. I ran 7 miles the day my mother refused to meet my girlfriend. I let myself cry, tears merging with sweat, drying in the chilly November wind. (Emily May Anderson, October 12, 2012, andirantumbler.com)

Women outpace men in participation in road races and recreational running in the USA as evidenced in popular online running sites such as *Women's Running* and *Runner's World*, and the emergence of self-made women's running blogs like *Fit and Feminist* (https://fitandfeminist.com/about/) and *Fat Girl Running* (http://fatgirlrunning-fatrunner.blogspot.com/). Many of the stories about women runners in popular online running sites and in blog posts show that women run to manage their physical and mental health; regular physical activity aids weight control, adds to positive stress relief and mental outlook, and may prevent premature death, cardiovascular disease, diabetes, and cancer (Brown, Burton, & Rowan, 2007; Martins, Morgan, & Truby, 2008; Szabo & Ábrahám, 2013; Warburton, Nicol, & Bredin, 2006). One downside to the women's running boom could be the marketplace. Jaclyn, one of the women I interviewed, felt that running has become commercialized. She enjoys the low-cost nature of running—"You just need a pair of shoes"—but talked about the expensive gear that online magazines sell, and the cost of training programs. She bought a few training programs previously (e.g., Training Peaks), but prefers Hal Higdon's programs because they are free. Rachel does not subscribe to running magazines, either. "I read them sometimes, but all they want to do is sell me products and tell me to lose weight. I'm so tired of the guilt-driven marketing. They don't give me what I want." Thus, as I have been arguing in this ethnography, the socio-cultural influences on women who run are more important than biology.

Running and Embodied Experience

Women's running bodies are embedded in cultural discourse about appropriate ways of being; the dominant cultural image of a woman runner and the normative running body is white, thin, straight, fast, feminine, middle-class, and disciplined (Hanold, 2010; Jutel, 2009). According to Running USA (2014), the typical US female runner is 39 years old, 140 pounds, 5 feet 5 inches tall, and has an average Body Mass Index (BMI) of 23.3; less than half of the women they surveyed were content with their weight (39%) and fitness level (41%). Women begin running primarily for exercise, weight control, and because of family and friends, and keep running to stay healthy and in shape, and to relieve stress. We need to see the bodies and stories from a variety of women from thin to average to fat, from recreational to fitness to competitive runner. In this chapter, I present an analysis of online material about running for women that shows how women's embodied experience matters.

My colleagues and I investigated how narratives of running in women's online running magazines subvert mainstream discourses of what being female and being active mean in terms of identity, motivation, and practice (Faulkner, Tetteh, & Behrman, 2016). We interrogated dominant understandings of women who run in two popular women's online running sites, *Zelle* and *Women's Running*, critiquing how health, appearance, and identity are constructed within dominant cultural representations (see the Appendix for method). Our findings suggest that women often run to combat everyday moments of difficulty because running is a space independent from obligations and expectations; running represents embodied physical and mental strength training for women runners.

In October 2014, *Zelle*, a new women's running website, was launched by *Runner's World* as "a site for women runners, by women runners." The editor of the site, Elizabeth Comeau, described *Zelle* as a counterpoint to larger cultural discourses about women's bodies, roles as mothers, and roles as athletes that circumscribe the experiences of women who run (Comeau, 2014a). *Zelle* was discontinued by *Runner's World* in December 2016, the woman-focused material now integrated into their email newsletters and on their website, and on June 7, 2017, Betty Wong Ortiz became the first female editor of *Runner's World* in its 51-year history. This suggests the prominence and permanence of women runners.

92 Women Running Online

I compare the content of *Zelle* and its competitor, *Women's Running*, through the concept of standpoint embodiment to understand how women experience their bodies, identities, and selves as runners (Davis, 1997; Velija, Mierzwinski, & Fortune, 2013). I discuss this work as well as my critique of selected blogs written by women runners.

Standpoint theory helps scholars examine experience through social construction recognizing how knowledge, including bodily knowledge, and our accounts of this knowledge are neither neutral nor universal (Collins, 1986; Harding, 2004). The body as it is lived and constructed through gender, sexuality, and other social positioning is important for feminist theorists to attend to and connect women's actual experiences in their bodies with the political project of deconstructing the mind–body split (Davis, 1997). In addition, I discuss my interviews with women runners (see Chapter 3) and select content from women's running blogs—*Run Like a Girl*, *261 Fearless®*, *Run Like a Mother®*, *Fit and Feminist*, *Fat Girl Running*—focused on real women's experiences running with their mother, fat, queer, and average bodies as another counterpoint to commercially produced material.

Examining embodiment for women runners meant including bodily knowledge from their standpoints and perspectives, and critiquing how dominant messages about women's bodies frame the experiences of women runners. The understandings of what it means to have a runner's body and to be a "real runner" may differ for those not situated as the recipients of privileged race, gender, and normative understandings (Collins, 1986). What we assume to be a neutral or universal perspective is often the standpoint of dominant groups and is reflected in the dominant discourse that limits or excludes those lived perspectives not affirmed within the dominant culture. Therefore, my analysis of online sources for women runners identifies how health, appearance, relationships, and identity are constructed in dominant cultural ideas of women who run and how they negotiate these meanings in online forums.

Women's Running Online

Women get information about physical exercise and nutrition from a variety of sources including books, magazines, family and friends, and health professionals (Clarke & Gross, 2004; Zawila, Steib & Hoogenboom, 2003). In Running USA's 2014 survey, women most often read *Runner's World*, *Fitness Magazine*, *Women's Health*, *Shape*, and *Women's Running*, and visited Active.com, Runnersworld.com, and MapMyRun.com online most often. About half of the women I interviewed told me they go online to read articles and advice about running, especially if they are training for a race.

> I read *Runner's World* things on Facebook. I look up a lot about different training schedules. When I trained for my first marathon, I read lots of articles online. There is a lot of information out there. Some of it is good. Some

of it is not. Once I experienced a marathon, I didn't need to read anymore. I had my own experiential journey. It is different for everyone. It is inspirational to read sometimes. Sometimes it just makes me feel guilty. (Jessica)

I asked my interviewees: *Do you ever look up information online about running? Do you subscribe to any running magazines? If yes, what do you think of the information?* Kjerstin said:

All of the time! I eat, sleep, and breathe running. The only magazine I get is *MN Runner* so I can keep up with local races and run news. I do love reading about running and frequently read things from *Runner's World, Competitor, Trail Runner Magazine, USATF*, and a lot of individual athletes and companies. Many times I think the info is geared towards less obsessed runners than myself, but I still like reading advice from different athletes and coaches.

The biggest thing I get out of running media is inspiration and motivation from seeing what other people are doing. I connect with other runners through Instagram and Strava in addition to reading articles online and reading books/magazines. There is nothing better to get you out of bed and running than seeing everyone else doing work and having run adventures!

There is a lot of conflicting advice and different ideas out there about right and wrong ways to run. I just like reading it all, filtering through it and hopefully doing some learning. I am lucky to have a degree in exercise science, a good coach, and a lot of life experience behind me so I don't follow crazy run fads like barefoot shoes or chia seed obsessions. But I enjoy reading about it all and taking in the bits and pieces that are based in solid science.

Many of the women I talked with also used online running coaches and running apps that helped them train and track their running such as *dailymile, runcoach, RunKeeper, daily activity, runtastic, map my run*, and *Garmin Connect*. Others used GPS watches, stopwatches, or an iPhone to track their mileage. Most posted their running on social media sites, including Facebook running groups. Shireen told me that she only uses social media with groups that she and her friends create, but talked about how that support was part of what she loved about running. Others went old school without the latest technology.

I don't have a smartphone or fancy Garmin watch. I have an iPod shuffle for music with my favorite fast-paced running playlist. I have a cheap digital watch from Walmart that has a stopwatch feature. I like to keep track of my minutes-per-mile as I go. I write my mile markers on my arm in pen so I can know my pace. The ink sweats off after about 10 miles and I can no longer read it. (Jessica)

Women told me that keeping track of their information and progress helped them to become runners.

> I use *RunKeeper* because I can map my routes and save them. I am able to find loops and know exactly where each mile marker is. After my run, I can input my pace and my time. I can look back at my old training runs from other times and see how much I improved my pace. When training for a full marathon, I use a different kind of approach (i.e., 3 days a week plan) than most training plans. I follow it exactly, because that's the kind of person I am. (Jessica)

Many women who run form a community with others who share similar aspirations, primarily through online forums, because of the support available on such platforms. This practice is becoming common as many magazines have online presence as a way of reaching and engaging their readers (Kaiser, 2001). These online platforms offer safe spaces for women to share their running experiences. For example, this focus on physical and mental strength through running is supported in online forums like *261 Fearless®*, Katherine Switzer's global running organization (www.261fearless.org/).

> It is the mission of 261 Fearless to bring active women together through a **global supportive social running community**—allowing fearless women to pass strength gained from running onto women who are facing challenges and hence sparking a revolution of empowerment. 261 is the symbol that unites us as empowered runners.

The organization creates opportunities for women to learn coaching, to support one another in running communities, and to find the fun and social aspects of running, and provides expert information on women's health, women's running, and women's running groups.

Running and Identity

The online narratives women runners present subvert the dominant cultural image of the white, thin, straight runner who runs solely to maintain aesthetics, and demonstrate how women run against idealized images of a "running body" (Hanold, 2010). Mavis said, "I see super toned women and fat women in the media. I never see women who are average, size 10 and size 12." Rachel told me that women runners in the media are "often very thin, very tan, very photo-shopped. They don't represent regular women runners. Most of us aren't super fit. We have normal bodies, not athlete bodies. I try to ignore media messages about women runners." As various cultural institutions frame female bodies in monolithic ways, the women on *Zelle* and in blogs often challenge these notions by presenting non–airbrushed, average sized women. Ultra-distance runner Mirna Valerio writes about running on her blog, *Fat Girl Running*: "People always say to me, 'Anyone who runs as much as you do deserves to be skinny.' Of course, what they're really saying: 'If you do all this running, why are you still so fat?'" (Brant, 2015, para.1). Women who run any distance, speed, and manner with *their bodies* are runners.

Zelle and *Women's Running* highlighted themes related to health and appearance. Pamela told me that she sometimes reads *Runner's World*, but she mentioned that the women who appear in the pages often look like "models in running clothes." She had mixed feelings about how women runners are portrayed in media. In some ways, she considered the depictions to be getting better in that there were representations of more average runners. She went on to tell me that she has been at an unhealthily low weight before; I assumed she was connecting the media images and her self-perceptions. Sarah was anorexic at 18, and found that running helped her to be present in her body and to appreciate what her strong running body could do. "My thighs love one another, they kiss each other. I don't want thigh gap." She said that the women runners she most often saw in media were not real. "Women who exercise are not Hollywood perfect." Kjerstin also had mixed experiences with the portrayals of women runners in the media.

> I think they can be amazing and also disappointing. It is fantastic when an athlete like Kara Goucher (my hometown hero) is on a cover or in the news. Athletes like her are beautiful people, women and athletes. I think the media does a good job of showing that, but it doesn't happen often enough. There are zero women's fitness/health magazines that actually feature women athletes of any kind. All of the focus is on looking skinny and losing weight—not on being a healthy, whole person. The women featured are almost always celebrities and rarely athletes. There are so many incredible role models in women's athletics and running, and I think it is terrible that they aren't given the media attention they deserve.

One idea present in *Women's Running* and *Zelle* was that women consider running as more of an identity versus solely as a way to be healthy and look good. They critiqued running as a means of achieving a normative "running body." *Women's Running*, though, often features weight-loss stories, which some runners, like Shireen, find inspiring. Other women featured on the websites did not describe their reasons for running in terms of the size or shape of their bodies; their bodies were not the defining factor in their running performance and experiences. Most of them run because they love the sport, because it is a way of being, and because running is part of their identity.

> The truth is running is a huge part of my identity. When I need to process my thoughts, I run. When I'm angry, I run. When I'm happy, I run. When I'm sad, I run. I've run to help get over old relationships, I've run to process

all the details about new ones and everything in between. (Eat, Pray, Run DC, 2015, para. 2).

Sometimes running isn't just about pace, miles, and personal bests. Sure, striving to go faster will always be a motivation to keep moving. But the main reason to run is out of love. Love for yourself, love for how it makes you feel, love for to run. Let go of the stress that surrounds the goals you have set for yourself this weekend. When you head out for your long run or line up at the start of a race, remember you are running because it makes you happy. And that is the quickest way towards success (Dietz, 2014, para. 1).

Running allows me to hold onto my own identity. It helps me to be more than happy to wear all of the many different hats required of me such as being a great parent, a good friend, an effective employee, an active citizen in my community or even a spouse again someday (Hungry Runner Girl, 2015, para. 5).

These examples demonstrate that many women runners do not reference their body sizes and shapes when they share their running experiences; they focus on their love for running. When issues related to body shape and size came up in online forums, women argued that a woman runner is not defined by the size of her body because the most important thing about running is not weight loss but the total running experience: "It's about the way you feel while running, not a number on the scale!" (Pattillo, 2014). Sebor (2014) echoed this view: "Today, I'm more than 30 pounds heavier—a figure that would have been a nightmare for my 14-year-old self. But I'm exponentially happier, I can run faster than ever and I've never felt more beautiful" (para. 5). The numbers on the scale do not tell women much about how running feels and what it means to them.

Looking back at my training log, I noted all of my recent PRs. I've impressed myself by the amount of miles I've been able to run … All of these things matter so much more to me than the three numbers on the scale. Strength and performance are much more valuable to me as an athlete than what the scale said. (Gattsek, 2014, para. 4)

The majority of women on the forum agreed with this point of view. For example: "I am an ultra-runner logging 60–70 miles per week and just finished my first 100-mile race. Not a pound thinner! So glad I am not alone! Cheers!" (Jenkins, 2014). Weight loss may not be the reason women keep running, as Steil (2014) mentioned: "I think this article shows a great way to find value in ourselves rather than in the number on the scale."

Women were not only supportive of plus-size bodies, they also supported thin bodies. Women runners critiqued how a woman runner with an athletic body is often referred to as a "man" in women's running circles. They rejected this shaming of women runners' bodies, noting how keeping quiet about sexist remarks harms not only the women in question, but all women. For instance, Trumbore (2014) said, "Fat-shaming and Thin-shaming and Any-Kind-of-Body-Shaming should be put on the same level as Racism and Hate Speech. It is not acceptable, no matter what. EVER. Thank you for being brave enough to post about it!" Also, women responded to news that some followers of *Runner's World* on Facebook criticized runners with athletic bodies as setting bad examples for young children. "Since when has calling a woman's body out, very publicly on social media, been OK? Oh, right. It always has been. But never would I have imagined members of the running community to be so terribly mean" (McGoldrick, 2014, para. 5).

Women runners also encouraged one another to look beyond the limitations of their physical bodies and focus instead on their inner strength. One woman said,

> I looked through pictures of someone I didn't recognize. What I imagined that I looked like, a strong, happy woman, was on the screen, in high resolution, looking like a sausage in too-tight casing, a look of sheer exhaustion on her face … I tried not to focus on the wideness of my thighs, but rather the strength in them that allowed my body to be upright and moving forward at a decent pace. I tried not to focus on the size of my arms, but rather the motion that goes into them to propel my body to the finish line. And I tried not to focus on the exhaustion in my face, but rather the fact that the emotion that I saw was so very real and honest for what I was going through at that time. (Birch–McMichael, 2014, para. 5).

Women runners did admit in these forums that their inner critics made body-shaming possible, because they had internalized the notion of the ideal running body.

> I have spent so much time focused on what others think of my body and it's crazy thinking. I am training for my first half in April that is what matters. Not the amount of dairy product on my thighs that jiggles as I run toward that high. Right? (Red3horn, 2014)

And they compared their own bodies against this ideal.

> I can totally relate. Lost 85 pounds running and working out regularly over the last two and a half years. Used to avoid putting on tank tops at all, but I recently realized that is no longer a problem and that they are actually my preferred workout top! What a difference a little time makes! (Comeau, 2014b)

98 Women Running Online

Zelle and *Women's Running* feature women who run for a variety of reasons beyond impression management in an effort to counter the pressure for the "ideal running body." Mirna Valerio, an African American runner, blogs about running and life for *Women's Health*.

> I am a big girl, a big runner. A fat runner. After all, the name of my blog is *Fat Girl Running*. In it I hope to spread the word that being a larger person and running aren't two mutually exclusive things or ideas and I hope that you will indulge me for a minute in exploring the idea of being a larger person and being fit.
>
> Society has its own ideas about how a runner should look. This of course, is sometimes accompanied by assumptions about a person's lifestyle, nutrition habits and overall health. Sometimes people look at me and make assumptions (and then comment aloud or in writing, like on the Interwebs) about how I must eat junk food all day. Or, that I probably spend a large portion of my day on the couch in front of the television. Or that I couldn't possibly run/jog the miles I do and still retain the body shape I live in. I'm not sure why this matters to anyone who isn't me. I'm not sure why I can't just run and live.
>
> I still have a ways to go in terms of reaching whatever my body determines its optimal fitness is, but I know that at present my blood sugar is and has been stable, my blood pressure is normal (after a short spike while I was writing my book—can we say MAJOR anxiety?), my cholesterol is within normal ranges for my age, and my C-reactive protein is on point. What does this mean? I will probably always be heavier than the norm, but as long as I continue running and engaging in the amount of physical activity that makes a significant difference in my metabolic profile, and as long as I maintain a pretty healthful lifestyle, I will likely be alright. And I'm okay with that. (Valerio, 2017)

However, *Women's Running*, in particular, continues to reinforce a thin-ideal stereotype. This is especially true for the nutrition section of the site; it seems that health is synonymous with model-thin, conventionally beautiful women. For example, Miller, Pilkington, and Sebor's (2014) article entitled "Our Top 4 Favorite Mid-Run Chews" features an image of a thin, white woman with long blonde hair and blue eyes. Her gaze meets ours while she sexily (and awkwardly) places a fudge-sized protein chew between her perfectly white teeth. It is also notable that, despite the title of the article, the woman appears clean, sweat-free, and perfectly manicured—traits that are unachievable while running.

Although this image was on the site in 2014, it is reminiscent of Jean Kilbourne's (1999) argument that women are often sex objects in advertising.

A woman is a sex object if her sexuality is used to sell something (Stankiewicz & Rosselli, 2008). In other words, women are not supposed to eat, but instead serve as ancillary *objects* that make food appear appetizing. *Women's Running* reinforces this objectification as the picture implies that mid-run chews can be used as objects for flirtation rather than an energy boost. The article, itself, is a paradox. The primary goal for the "mid-run chew" is to maintain energy throughout one's run, yet the authors offer the caloric content of each chew, suggesting that the best chew is the one lowest in calories. But here's the thing, runners can't burn energy without calories.

What is most incongruous on the *Women's Running* website is the overarching idea that women should be ashamed of eating unhealthy food; *all* of the articles that mentioned non-healthy food described how eating them makes one "guilty" or "sinful." For example, Hynes' (2014) article on "3 Tips for Breaking the Holiday Bloat" reinforces the idea of guilt and a necessary atonement through dieting: "With the holidays in full swing, you're probably finding yourself in a constant state of bloat. Ugh. You wake up feeling guilty and gross." Hynes suggests doing a detox based on the crash diet phenomenon, the Maple Syrup Diet, which encourages participants to drink *only* a mixture of lemon water, cayenne pepper, and maple syrup. The emphasis on bingeing and purging is a consistent theme throughout the site.

Binging and purging habits are harmful for women's health. A binge/purge cycle entails consuming an atypical amount of food and then ameliorating it through extreme exercise, self-induced vomiting, use of laxatives, or extreme dieting. Yet despite *Zelle* and *Women's Running* push for health, the nutrition pages regularly encourage unhealthy eating behavior. Furthermore, readers appear to value articles that focus on "dieting" or "cleansing." The most shared and pinned nutrition article between September and December 2014 on *Women's Running* was a day-by-day diet menu, "The Athlete's Cleanse" (Odweller, 2014). The article offers a 3-day diet for runners that endorses the "cleanse" phenomenon. Odweller suggests doing the cleanse after your post-race meal splurge, and positions the cleanse as a redemptive act for eating a "splurge" meal. The concern is the focus on "cleansing" the system rather than focusing on energizing the body for activity; the diet plan that Odweller suggests is below the normal caloric count that a *non-runner* would consume and significantly below the caloric content that a runner would consume. This reinforces a binge/purge diet approach.

These articles on *Women's Running* and *Zelle* tell a story about the need to discipline the female body. Although most women on *Zelle* embrace their bodies, the images on the nutrition page are part of the discipline of women's bodies and are emblematic of a long history of crash diets presented to women in household magazines. Running, exercise, and fitness are overshadowed by diets that encourage weight loss for appearance rather than weight loss for health.

Running and Motherhood

For some women, running and motherhood are equally important aspects of their lives and identities. Women I talked with told me how running helps them become better mothers, partners, and people; motherhood and running are compatible pursuits. In online forums, women expressed similar ideas.

> Running, with its demands that I mentally focus and take care of my body, ultimately made my transition to motherhood easier. And becoming a parent improved my running, requiring that I make it a priority and get the most out of each workout in the limited time I had (Williams, 2014, para. 8)

> As a mom of two boys under 4, my days are packed. I'm much busier than I was 10 years ago. But I've gotten 100x better at making time or finding those tiny windows of time where I can run. Running is no longer something I fit in when I can. I now fit in other things around my daily run (NYC Running Mama, 2015, para. 10).

> In as many years as I've been running, no runs have brought me more joy than those with one or both of my kids alongside me. Whether it's a local 5K or just doing a mile around the neighborhood, these are the footsteps I treasure most (Loudin, 2014, para. 1).

> While I was pregnant, I ran about 1,200 miles with my little girl. I spent those miles thinking of her, imagining her, deciding what type of mom I wanted to be and praying for her. There were many times that I would just start crying as I was running (those pregnancy hormones really got me) because I was so excited to meet her. Running also gave me confidence during my pregnancy. It allowed me to feel like myself again as I was experiencing my body changing in so many different ways (Hungry Runner Girl, 2014, para. 3).

However, combining these two roles is not usually easy: "I had visions of a smiling child and happy mom logging endless miles together. But the reality was oh-so-different" (NYC Running Mama, 2014, para 1). Alex found the ubiquitous stories about how women runners take time away from their children to train annoying. "If a man has children, he should be involved and would be leaving to train, so why talk about women?" Some women expressed guilt when they leave their children behind to go for runs.

> If I'm out on a long training run without my son … I spend nearly my entire run thinking about how I need to run faster so I can get home and build that fort we made plans to create out of couch cushions … I hope, in

some way, my son understands that I'm doing the best I can for myself, and for him (Comeau, 2014c, paras. 5, 13).

In the comments to this post, a few women wrote that they also felt guilty when they ran without their children, but other women did not. One woman left a comment about feeling guilty for a night out with friends and her husband, but not when she goes for a run without her children because she is setting an example of an active, fit mother for them: "Moms have enough to beat themselves up about, don't beat yourself up over being healthy and active, showing your children what a healthy/ active person looks like, and teaching them the importance of family/work/health/ management" (Momma's Gotta Run, 2014). These women see taking time away from their children to run as modeling healthy behavior. Mavis felt frustrated that women runners were often portrayed with their children in strollers in the media. She told me that being a mom and wife were just roles she played, but being a runner was part of who she was. The online running site and Facebook group, Run Like a Mother®, offers resources for women runners who mother to combat the attitude that one can be a mother OR a runner. There are also women–only running events and web pages devoted to empowering girls and women to run. For instance, *Run Like a Girl* offers women–only running events and running events for children to show that being a mother and being a runner are not mutually exclusive.

Mind/Body to MindBody

Examining online sources about women running contributes to an embodied feminist project and speaks to the call for a more physical feminism (McCaughey, 1997; Velija et al., 2013). We see this in Facebook groups and online running groups that support women runners' empowerment (e.g., Fit and Feminist; Fit Is a Feminist Issue; 261 Fearless®, Run Like a Mother®). Running develops women's physical and mental strength enabling us to embrace the contradictions between a deconstructed focus on the mind/body split and to focus on our "actual material bodies ... their everyday interactions ... and [how we experience] through our bodies with the world around us" (Davis, 1997, p. 15). Running is a feminist issue, and running is a feminist act. I argue that women who run use running as a way to combat everyday moments of difficulty as running forges a location independent from relational, familial, and employment obligations. The alternative narratives of gendered embodiment displayed in online women's running sites contest the false dialectic of public and private spaces and demonstrate a feminist physical embodiment (Baxter, 2011; Davis, 1997; Orr, 2006). That is, the act of running centers women in their bodies as active agents who can avow their own identities as runners across all social categories.

When comparing *Zelle* and *Women's Running* with blogs, there were instances of empowerment and disempowerment. As various cultural institutions frame female bodies in monolithic ways, the women on *Zelle* and in running blogs often challenge these notions by presenting non-airbrushed, average sized women, countering the dominant ideal of the female disciplined body and the meaning of health as rooted in appearance and materialism. Most sites, though, advertised expensive running gear and travel without discussion of running as one of the least expensive sports, confirming some women's beliefs that running is becoming commercialized. Embodiment for women who run means self-determination, their choice of identities, and an integration of health and running as women react against the image of the dominant female running body. Running offers women an important marker of embodied identity in the everyday moments of difficulty related to work, relationships, and health, and many online sources now support women's embodied running selves with writing, images, and sounds from our various perspectives.

References

Baxter, L. (2011). *Voicing relationships: A dialogic perspective.* Thousand Oaks, CA: Sage.

Birch-McMichael, M. (2014, November) Why race photos don't show the whole picture. Zelle. Retrieved from http://zelle.runnersworld.com/chatter/why-race-photos-dont-show-the-whole-picture

Brown, W. J., Burton, N. W., & Rowan, P. J. (2007). Updating the evidence on physical activity and health in women. *American Journal of Preventative Medicine, 33*(5), 404–411.

Brant, J. (2015, July 21). Is it possible to be fat *and* fit? At 250 pounds, distance runner Mirna Valerio provides an inspiring example. *Runner's World.* Retrieved from www.runnersworld.com/runners-stories/ultra

Clarke, P. E., & Gross, H. (2004). Women's behaviour, beliefs and information sources about physical exercise in pregnancy. *Midwifery, 20*, 133–141.

Collins, P. H. (1986). Learning from the outsider within: The sociological significance of Black feminist thought. *Social Problems, 33*(6), S14–S32. Retrieved from http://ucpress-journals.com/journal.asp?jsp

Comeau, E. (2014a). #What the Zelle. *Zelle.* Retrieved from http://zelle.runnersworld.com/chatter/what-the-zelle

Comeau, E. (2014b, November). Why my running body image is best on ice. *Zelle.* Retrieved from http://zelle.runnersworld.com/chatter/why-my-running-body-image-is-best-on-ice

Comeau, E. (2014c, October). Marathon mom guilt. *Zelle.* Retrieved from www.runnersworld.com/chatter/marathon-mom-guilt

Davis, K. (1997). Embodying theory: Beyond modernist and postmodernist readings of the body. In K. Davis (Ed.), *Embodied practices: Feminist perspectives on the body* (pp. 1–26). London: Sage.

Dietz, K. (2014). Chasing happiness. *Women's Running.* Retrieved from http://womensrunning.competitor.com/2014/10/inspiration/friday-photo-inspiration-chasing-happiness_31821

Eat, Pray, Run DC (2015, January). What it means to be a runner. *Women's Running.* Retrieved from http://womensrunning.competitor.com/2015/01/eat-pray-run-dc/eat-pray-run-dc-means-runner_34685

Faulkner, S. L., Tetteh, D. A., & Behrmann, E. M. (2016). Women who run: Identity, embodied experience, and gendered expectations in popular women's running websites.

Paper presentation at the annual Central States Communication Association. Grand Rapids, MI.

Fit and Feminist (n.d.). Retrieved from http://fitandfeminist.wordpress.com

Gattsek, S. (2014, September). Weight matters: On tossing the scale. *Runner's World*. Retrieved from www.runnersworld.com/chatter/weight-matters-on-tossing-the-scale

Hanold, M. (2010). Beyond the marathon: (De)Construction of female ultrarunning bodies. *Sociology of Sports Journal, 27*, 160–177.

Harding, S. (2004). Introduction: Standpoint theory as a site of political, philosophic, and scientific debate. In S. Harding (Ed.), *The feminist standpoint theory reader: Intellectual and political controversies* (pp. 1–15). New York, NY: Routledge.

Hungry Runner Girl (2014, November). Why I loved running during pregnancy and tips for pregnant runners. *Women's Running*. Retrieved from http://womensrunning.competitor.com/2014/11/hungry-runner-girl/hungry-runner-girl-loved-running-pregnancy-tips-pregnant-runners_32376

Hungry Runner Girl (2015, January). Making time for yourself. *Women's Running*. Retrieved from http://womensrunning.competitor.com/2015/01/hungry-runner-girl/hungry-runner-girl-making-time_34272

Hynes, K. C. (2014, December 11). 3 tips for breaking the holiday bloat. *Women's Running*. Retrieved from http://womensrunning.competitor.com/2014/12/nutrition/3-tips-breaking-holiday-bloat_33758

Jenkins, K. (2014, November) On tossing the scale [Web log comment]. Retrieved from www.runnersworld.com/chatter/weight-matters-on-tossing-the-scale

Jutel, A. (2009). Running like a girl: Women's running books and the paradox of tradition. *Journal of Popular Culture, 42*(6), 1004–1022. doi:10.1111/j.1540-5931.2009.00719.x

Kaiser, U. (2001). The effects of website provision on the demand for German women's magazines. *ZEW Discussion Papers*, No. 01-69. Retrieved from www.econstor.eu/bitstream/10419/24490/1/dp0169.pdf

Kilbourne, J. (1999). *Killing us softly 3* [DVD]. Northampton, MA: Media Education Foundation

Loudin, A. (2014, November). Building her own running team. *Zelle*. Retrieved from http://zelle.runnersworld.com/chatter/building-her-own-running-team

McCaughey, M. (1997). *Real knockouts: The physical feminism of women's self-defence*. New York, NY: New York City Press.

McGoldrick, H. (2014). What a real runner looks like? *Zelle*. Retrieved from http://zelle.runnersworld.com/chatter/what-a-real-runner-looks-like

Martins, C., Morgan, L., & Truby, H. (2008). A review of the effects of exercise on appetite regulation: An obesity perspective. *International Journal of Obesity, 32*, 1337–1347.

Miller, N., Pilkington, C., & Sebor, J. (2014, September 6). Our top 4 favorite mid-run chews. *Women's Running*. Retrieved from http://womensrunning.competitor.com/2014/09/nutrition/4-favorite_29596

Momma's Gotta Run (2014, November). Marathon mom guilt [Web log comment]. Retrieved from www.runnersworld.com/chatter/marathon-mom-guilt

NYC Running Mama (2014, September). Stroller running tips. *Women's Running*. Retrieved from http://womensrunning.competitor.com/2014/09/photos/nyc-running-mama-stroller-running-tips_29640

NYC Running Mama (2015, January). Does running ever get easier? *Women's Running*. Retrieved from http://womensrunning.competitor.com/2015/01/nyc-running-mama/nyc-running-mama-running-ever-get-easier_34262

Odweller, L. (2014, October 7). The athlete's cleanse. *Women's Running*. Retrieved from http://womensrunning.competitor.com/2014/10/nutrition/recipes/athletes-cleanse_31101

Orr, D. (2006). (Ed.). *Belief, bodies and being: Feminist reflections on embodiment*. Lanham, MD: Rowman & Littlefield.

Pattillo, A. (2014, October). Racing weight. *Women's Running*. http://womensrunning. competitor.com/2014/10/just-for-fun/run-numbers-racing-weight_31199

Red3horn (2014, November). Why my body image is best on ice [Web log comment]. Retrieved from www.runnersworld.com/chatter/why-my-running-body-image-is-best-on-ice

Running USA (2014). 2014 Women's national runner survey. Retrieved from www. RunningUSA.org.

Run Like a Mother. Retrieved from www.runlikeamother.com/

Sebor, J. (2015, September). A stronger runner. *Women's Running*. Retrieved from http://womensrunning.competitor.com/2014/09/editors-corner/editors-corner-a-stronger-runner_29555

Stankiewicz, J., & Rosselli, F. (2008). Women as sex objects and victims in print advertisements. *Sex Roles*, *58*(7/8), 579–589. doi:10.1007/s11199-007-9359-1

Steil, R. (2014, November). On tossing the scale [Web log comment]. Retrieved from www.runnersworld.com/chatter/weight-matters-on-tossing-the-scale

Szabo, A., & Ábrahám, J. (2013). The psychological benefits of recreational running: A field study. *Psychology, Health & Medicine*, *18*(3), 251–261. doi: D10.1080/13548506.2012.701755

Trumbore, J. (2014, October). What a real runner looks like [Web log comment]. Retrieved from www.runnersworld.com/chatter/what-a-real-runner-looks-like

Valerio, M. (2017, February). Why health is more complicated than BMI. *Runners World*. Retrieved from http://womensrunning.competitor.com/2017/02/fat-girl-running/health-weight-more-complicated-bmi_72409

Velija, P., Mierzwinski, M., & Fortune, L. (2013). "It made me feel powerful": Women's gendered embodiment and physical empowerment in the martial arts. *Leisure Studies*, *32*(5), 524–541. doi:10.1080/02614367.2012.696128

Warburton, D. E. R., Nicol, C. W., & Bredin, S. S. D. (2006). Health benefits of physical activity: The evidence. *Canadian Medical Association Journal*, *174*(6), 801–809.

Williams, M. H. (2014, December) Why she runs: To adapt to motherhood. *Zelle*. Retrieved from http://zelle.runnersworld.com/chatter/why-she-runs-to-adapt-to-motherhood

Zawila, L. G., Steib, C. M., & Hoogenboom, B. (2003). The female collegiate cross-country runner: Nutritional knowledge and attitudes. *Journal of Athletic Training*, *38*(1), 67–74.

5

RUNNING AS FEMINIST EMBODIMENT

Winter Run

I wish that I were this winter's bitch
as she smacks my cheeks to a rosacea red
slides jagged fingers under my shirt until I slip,
my mouth all stretched circle and tongue, agape.
She kisses my tendonitis with ice knuckles
so I cry a polar vortex of popsicle tears
that she slaps across my lashes like a brass buckle.
Though her morning breath abrades my corneas
I damn the counties snow emergencies,
double-knot my laces and swerve around downed cars
stretching and cracking plantar to please.
I know that I am this winter's bitch
as I skid, a pile of yellowed snow in a ditch. (Faulkner, 2014)

Writing and Running

Both running and writing are highly addictive activities. (Oates, 2003, p. 32)

★★★

Writing Practice, May 2016

Wake up at the usual time. Roll over like you are still asleep, so the dog won't hear.
You just need another minute. Curse the minute that transformed into an hour.

106 Running as Feminist Embodiment

Damn the semester's night class that took your will for early morning writing espresso and listening to your house's predawn snoring. No free hour of writing for you today as the nagging obligations of domesticity and paid work crouch at your writing desk littered with unpaid bills and a smear of dust. Forget how last week a student asked to see this magical writing space you describe in a *published* interview about your writing practice. Know that there is no magic. Hope that extra sleep was worth it. The writing practice you just slept through.

Recommit to your deadlines as you let the dogs rouse the snoring family. Deadlines as writing prompts usually get you to put words in a .doc file, to write at the desk tucked under the loft in your church house. Think about adopting your colleague's method of 500 words a day. Remember that never worked for you, and re-metaphor deadlines as the skeins of writing motivation that you knit into habit.

Instant message your writing buddy for the weekly check-in after you let the dogs out to pee. Complain about the meetings about meetings that you *let* interrupt your writing. Laugh when she suggests your grumpiness could make a good poem. Agree to exchange those essays—or whatever they are—her poetry about sex and desire versus the daily tedium of relationships in middle age, and your satire and screed about hating guns in a pro-NRA family. Yes, the ones that you both worked on when you were supposed to be writing something else. Feel thankful that you found a way to write about this and grateful for your writing buddy. Let yourself feel nervous about sharing, but maybe, just maybe, this piece might work. Make another date to talk about what you can do with that writing. List the projects you want to work on this month on a sticky note. Use the crayon you find on the floor by the dog bed since someone borrowed all of the pens that are supposed to be in the chipped cup by your desk.

Ignore your spouse wrangling the first grader: *Eat your fish sticks. Put on your clothes. Put on your clothes! No, you can't watch TV. We have to leave. Now. Now!* Find the overdue library book wedged between the dog and the couch cushion. Remind them to take the book back to school. Check the homework folder and cram it into the backpack. Is that smell pee? Clean the dog's irritation at some slight you most likely are responsible for off of your daughter's backpack, and notice that was puke you just stepped in on the carpet by the door. Cringe at the feel of bare feet on squishy moist fluids. Ask your new rescue dog, the dog now called Barfy, to stop eating all the nasty treasures she finds *outside* and leaving them *inside* for you. You really do not need such elaborate presents. Promise the dogs you will take them for a walk after you drink one more cup of coffee. And find your pants.

Eat the cold crust of cinnamon toast off your daughter's plate as you search for the orthotics for your running shoes. This counts as breakfast. Set the plate down for the dogs to eat the egg scraps. No need to scrape and rinse. You can work on writing as you run. Didn't you read some neuroscience article on the connection between problem solving, creativity, and relaxation? Didn't one of your favorite writers, Haruki Murakami, write, *What I talk about when I talk about running*, a memoir about running AND writing? The connection between exercise practice and writing discipline? You can write *Poets who run: how to write haikus in your head as you train*. Today. On this run you need to take because you are training for a half marathon. (Of course, you are always training for a half marathon.) Argue that this run will make you nicer in your afternoon meetings. Smile at your consideration of others. Email your student to say you will be 15 minutes late to your meeting.

Switch your usual route to avoid running by the Dean's house. Run in the depressing tree streets, the ones that are named for trees that do not exist in this hell of pavement and sprayed-on lawns with little signs that warn you no kids or dogs: Poison. Notice there are no squirrels here, either. Get lost in your resentment. Be mad when you actually *do* get lost in the wasteland because you were writing a poem in your mind about how the maze of depression in the rows of beige vinyl siding with a splash of brick disorients you.

You have to pee now. Don't hold your urine. Find the bulldozer parked at the end of new construction on Sequoia Street and enjoy squatting behind it. Realize that this is going to be a long run. How the heck do you get out of this development? Moan about being hungry now. Panic. Stop because you emailed the student. Praise yourself for getting in a long run in the middle of the week. Wonder if you will have time to take a shower before going to the office once you finally exit your middle-class suburban nightmare. It will be 2 more miles until you reach a shower.

Think about the reviews you agreed to write. Groan at the emails you will have to answer. Remind yourself to cancel the meeting about the meetings. You are sick. Yes, sick of overcommitting to things that do not help your writing. What if you over committed to writing? Ask if that is even possible. Decide that you will take Wednesday to write. And keep your meeting-free Friday writing day, too. Craft the emails on the last 2 miles you can now sprint.

Write a narrative poem about writing using lots of imperatives because you must give yourself permission to write. Remember a quote, one you uttered during an interview about being a mother who writes: "Be kind to yourself. Be impatient with the difficulty of finding time to write and don't quit."

Be impatient. Give yourself permission to write. Write down your plan with a sharpie marker. Tell the others. Do it. Do it with wet hair (Faulkner, 2017a).

★★★

> What I believe running and writing have most in common, at least for me, is the state of vulnerability they leave you in. Both require bravery, audacity, a belief in one's own abilities, and a willingness to live the clichés: to put it on the line, to dig deep, to go for it. You have to believe in the "it," and have to believe, too, that you are worthy. (Toor, 2014, para. 3)

My running practice is tied to my writing practice, so it is not shocking that my ethnographic analysis of women and running is impossible to talk about without talking about running *and* writing. Writing is not a disembodied activity. Writing about running proved to be an embodied experience; I worked out structural, content, and theoretical issues as I ran. And my running became a problem to work out in my writing. The prose poem I begin this chapter with speaks to my use of poetic inquiry as an analysis technique (Faulkner, 2017a). I can't pinpoint an exact moment when I recognized that poetic inquiry was the key to organizing and demonstrating this project as a feminist embodied ethnography. Most likely, the pieces were sweat out, sorted, and rearranged during runs. The problems of writing and running were tangled together for me in this project.

> **Writing Problems**: *How do I make running interesting to non-runners? How do I connect women's embodied experiences to the idea of running as a feminist act?*
> **Running Problems**: *How do I keep running? How do I keep running despite my maladies?*

When I ran, I thought about writing. When I wrote, I wanted to be out running. As I ran and as I wrote, I repeated a mantra that Haruki Murakami (2008), one of my favorite writers, discussed when talking about running:

> Pain is inevitable. Suffering is optional. Say you're running and you start to think, *Man this hurts, I can't take it anymore.* The *hurt* part is an unavoidable reality, but whether or not you can stand anymore it's up to the runner himself [sic]. This pretty much sums up the most important aspect of marathon running. (p. vii, emphasis in original)

I feel that this sums up the writing process, too. *Pain is inevitable. Suffering is optional.*

I often take a run when I'm stuck in my writing. Running enhances our creativity through stress reduction, focus, and efficiency, and our subconscious making connections (Presland, 2016). The movement minds and reminds me that writing is a process; I often find connections between seemingly disparate ideas, work out and create new problems, and write good lines in my mind as I run. There are good and bad runs. There are good and bad writing sessions. The writer and runner Joyce Carol Oates (2003) also uses running as a way to work out writing problems:

> The structural problems I set for myself in writing, in a long, snarled, frustrating and sometimes despairing morning of work, for instance, I can usually unsnarl by running in the afternoon. On days when I can't run, I don't feel "myself" and whoever the "self" is I do feel, I don't like nearly so much as the other. And the writing remains snarled in endless revisions. (p. 30)

Running as Feminist Embodiment **109**

I appreciate writers who run, as I recognize my process in their words. Training and writing practice turn motivation into habit. Oates (2003) considers running a vital part of the creative process. Running and writing are tied in our consciousness:

> Running! If there's any activity happier, more exhilarating, more nourishing to the imagination, I can't think what it might be. In running, the mind flies with the body; the mysterious efflorescence of language seems to pulse in the brain, in rhythm with our feet and in the swinging of our arms. (p. 29)

Of course, sometimes I run to get away from my writing. The neuroscience on how running helps our emotional regulation and the ties of running to creativity show the line between science and art is indeed thin and permeable (Bernstein & McNally, 2015). Evidently, the aerobic activity of running grows brain cells (Creer, Romberg, Saksida, Praag, & Bussey, 2010). Murakami (2008) talked about the connection between running and writing in his memoir. His use of running as solitude and as a means to become stronger physically and emotionally reflects my running practice. His belief that "running is both exercise and a metaphor" (p. 10) speaks to the connection I make between running and writing.

> People sometimes sneer at those who run every day, claiming they'll go to any length to live longer. But don't think that's the reason most people run. Most runners run not because they want to live longer, but because they want to live life to the fullest. If you're going to while away the years, it's far better to live them with clear goals and fully alive than in a fog, and I believe running helps you to do that. Exerting yourself to the fullest within your individual limits: that's the essence of running, and a metaphor for life—and for me, for writing as whole. I believe many runners would agree. (Murakami, 2008, pp. 82–83)

This is why as I was writing this chapter, I became distraught after being forced to take a 2-month running hiatus because of patellar tendonitis and a tight IT (iliotibial) band. If "running/writing keep the writer reasonably sane, and with the hope, however illusory and temporary, of control," then I was doomed (Oates, 2003, p. 36). I ran with knee and thigh pain for months, until my partner commented on how the knot on my right knee looked like a marble after I ran. The ice packs didn't help. *How can I write about running if I can't run?* This is why it took me months to admit that I was injured and to make appointments with the appropriate personnel: The chiropractor adjusted my hip over the course of several weeks, and my IT band was wound so tightly the massage therapist had to place her foot on the wall to get enough leverage to stretch it out. The orthopedist told me that my inner thigh was weak, and as a result was pushing my knee cap off-center causing my tendonitis. "This is what happens between 40 and 50," he said.

110 Running as Feminist Embodiment

I felt stuck. Stuck at 45 years old, between 40 and 50, between writing and not writing.

But then I did what I do best; I made a plan for my writing and for my healing. I took walks and bike rides and rowed and stepped miles on the gym machines after I lifted weights. I reframed the gym equipment as *not boring*. Every week, I would take a run to test out my progress and my knee would swell up, so I cancelled my plans to run the Glass City half Marathon one month before the race. I abandoned the online training plan I had paid for to get me to my 2:10:00 half-marathon goal. Wistfully, I watched others run, wanting to be running. I admired how a local woman, who I guess to be in her 70s, ran around the track in the gym and moved outside to the city park when it was warmer, her hair always impeccably coiffed. If I wanted to keep running, I had to rest. If I wanted to keep writing, I had to find other ways to move.

And then the Friday before the Glass City half, my BFF in Bowling Green messaged me to ask when I was going to pick up my race number. Her husband, also a dear friend, had signed up to run the race and DID NOT train. But, he was running it. I forgot that we had talked about running the race together.

"I'm not running. My last long run was a month ago when the 8 miles I ran made my knee burn," I typed.

"May I use your number? I can walk that far," she wrote. He was going to do the race. I admired that former athlete bravado. *Why not just pick up a race number and run 13.1 miles?* She was worried he would injure himself.

"I'll do it," I typed.

"*Sigh*," I thought. Friends run with friends. I message Marne, my running buddy, who is great at making running plans to see if what I wanted to do made sense. We would run/walk the race, so that he would not injure himself. Marne affirmed I would be fine and approved my plan. I ran/walked the only marathon I ever ran, and it worked. I would run the race with my friend, because that is what runners do. We have each other's back.

We do a 3-minute run/1-minute walk cycle until the last 3 miles of the race. Then we just run. My knee started burning after mile 3, so I am running injured. I know that it will hurt anyway, so I sprint the last 2 miles with the hope I can teach him a lesson about training. I still have energy to burn and glance down at my Garmin to see a 9-minute-a-mile pace. I hear him puffing behind me, but I need to be a runner, to remind myself that I can run. We finish in 2:36:33—12 minute miles. I am pleased and a little irritated. Someone who did not train should not have finished that fast. *But, wow, we did it together.* My knee is on fire, so I cash in my two beer tickets, something I typically don't do, and drink cold beer from the Glass City mug I got after the race finish.

My running buddy and friend, Marne, likened my healing to somatic memory.

> It kinda sounds like in writing of running, you're healing by running—like you're tapping into the … rhythm, flow, energy of it … that is something we

experience/feel as embodied directly in running. But we can even somatically tap into it even if we aren't physically running … like a present memory of touching that source that we are introduced to most viscerally in running, but really it's the door opening to what was already there. (Personal communication, June 1, 2017)

Once I connected my writing to any kind of movement, I felt better, physically and emotionally. Recognizing that I was a runner even if I was not running helped me see that I am a writer and an ethnographer. Like the four seasons, I was harvesting this ethnography and my running from the fall of fieldwork and shivering in the icy winter of needed rest until I could write and run in the green light of early spring (González, 2000). I began to listen to the interviews of the women runners I talked with on my iPod when I went for runs. This was another layer of analysis and commitment to running and writing. I liked the idea of running with my interviewees.

Running helps my writing and writing helps my running. In an interview for *Runner's World* in 2005, Murakami connected his running and writing process: "Without a solid base of physical strength, you can't accomplish anything very intricate or demanding. That's my belief. If I did not keep running, I think my writing would be very different from what it is now" (Lee, 2005, para. 23). Rachel Toor (2014) also compared writing and running to argue that the difficulty of training for long distance races mirrors the process of writing a book: both seem impossible at times.

> Running has made me a more disciplined writer, and writing has reminded me to be brave when racing. I've learned—I'm trying to learn—to keep faith in the face of flagging mind, body, spirit, and confidence. I know that any valuable achievement will require that I make myself uncomfortable and may well hurt.
>
> I've learned to recognize the pain: "Here we are again. This is the part that sucks. This is the place where I want to give up." (Toor, 2014, para. 15)

Writing an ethnography about women and running without the running and writing practices I had, would have made this project different.

> There is never enough of integrating the body into all the diverse ways of writing. The body does not want to be bracketed, or just be utilized as a semi-colon. The body wants to be a comma, constantly breaking up every little intention and action. (Snowber, 2016, p. 11).

In fact, it is precisely my running and writing practices that produced this ethnography. The writer, journalist, and runner Catriona Menzies-Pike (2017) wrote a memoir about running from a feminist perspective, using her background in

112 Running as Feminist Embodiment

literature. She made the point that writing about running means writing about something other than running, which mirrors Tony Earley's idea that good writing means writing about the thing and the *other thing*. "Every story is about the thing and the other thing" (Poets & Writers, 2016).

> Stories about running are often like this, in that they're about something else. They are tales of shape shifting, of the desire to shed one skin and step into another. One running story may be a parable on persistence or denial; another a warning. (Menzies-Pike, 2017, p. 5)

Thus, like these writers assert, *Real Women Run* is about women running, but it is also about identities in motion, the inseparable mind–body connection, and running as solitude, physical and emotional strength, and community.

Poetic Inquiry as Embodied Analysis

I add my personal story about feminism, running, and queer identity through poetic inquiry to this ethnography (Faulkner, 2009, 2017). I used poetic inquiry as a feminist ethnographer to contend with dominant discourses about women's running bodies, identities, and gendered/sexualized roles and the concomitant evaluations by acknowledging, examining, and altering the complex reality that women who run are not all Kara Goucher clones.

> "Poetic inquiry" is the use of poetry crafted from research endeavors, either before project analysis, as a project analysis, and/or poetry that is part of or that constitutes an entire research project. The key feature of poetic inquiry is the use of poetry as/in/for inquiry. (Faulkner, 2017, p. 210)

I consider the personal writing here to be more taboo than investigative writing because of the refusal to engage the mind/body split. A narrative poetic inquiry, in particular, is a form uniquely situated to re-present such liminal space.

> Poetry in research is a way to tap into universality and radical subjectivity; the poet uses personal experience and research to create something from the particular, which becomes universal when the audience relates to, embodies, and/or experiences the work as if it were their own. (Faulkner, 2017, p. 210)

In this ethnography I used poetry, poetic transcription of interviews, and poetic analysis of field notes to engage in embodied analysis. The poetry that I offer throughout this work can be considered ethnographic poetry (Faulkner, 2017). Using poetry in an ethnographic project is a way to demonstrate anthropological insights, to tell a story about fieldwork through the telling, retelling, and framing of embodied experiences with a poetic sensibility.

Running as Poetic Practice: Write Yourself as You Are with Purpose—Feminism and Poetry

> Knowledge to and from the body comes via practice. (Snowber, 2016, p. 8)

My interest in an in-depth examination of women's embodied running experience speaks to the call for a more physical feminism as articulated by McCaughey (1997) and Velija, Mierzwinski, and Fortune (2013). Ellingson (2017) argued that we ethnographers "expand our understanding of ethnography by considering how we do embodiment" and how "our participants do embodiment" (p. 81). I see women's physical and mental strength developed through running as a way to embrace the contradictions between a deconstructed focus on the mind/body split and the focus on "individuals' actual material bodies … their everyday interactions with their bodies and through their bodies with the world around them" (Davis, 1997, p. 15). The idea of "theories of the flesh" was integral to studying embodiment. Women of color have been using experiential knowledge or theories of the flesh to show the importance of storytelling and narrative in the representation of knowledge and everyday experience (Collins, 2000; Moraga & Anzaldúa, 1981). Further, the use of poetry is one means to theorize using the body.

> You can theorize through fiction and poetry; it's just harder. It's an unconscious kind of concept. Instead of coming in through the head with the intellectual concept, you come in through the backdoor with the feeling, the emotion, the experience. But if you start reflecting on that experience you can come back to the theory. (Moraga & Anzaldúa, 1981, p. 263)

Thus, I used an embodied approach that honors experience, voice, and the body to examine the idea of embodiment in women's running.

As I wrote in Chapter 2, "I'm doing participant-observation research on women's embodied experience, so I have to run, right? Running is research. Running as research." I used an embodied methodology in this ethnography. This meant the use of poetry as research analysis. "Poetry promises to return researchers back to the body in order to demonstrate how our theories arise out of embodied experience" (Faulkner, 2017, p. 214). This meant paying attention to "the smells and textures and bodily movements" in running to shift "away from discourse … to include more emphasis on materiality" (Ellingson, 2017, p. 83). An embodied ethnography defies the mind–body split; a feminist ethnography pays attention to the material and the discursive by taking up emotional, physical, and ideological space.

Standpoint Embodiment

The women's stories of running presented in this work demonstrate the idea of a physical feminism (McCaughey, 1997). Running is a way to build physical and emotional strength to challenge and resist normative running bodies, typical femininity, and staid expectations. There is a connection between physicality and consciousness. "While much of feminist thought had focused on how men and women come to incorporate sexist ideologies into their psyches, corporeal feminism insists on examining the ways such ideologies become inscribed and contested at the level of the body" (McCaughey, 1997, p. xii). We see how women run toward and run away from expectations, relationships, and ways of being.

> The impact of "running towards, or running away from" interactions, individuals, relationships, and/or institutions is how I answer the question, "Why do you run so much?" They too run towards and away from these things. Connecting literal and metaphoric meanings helps parallel the ideological understanding with the physical embodiment. That is, you/we literally run and embody that performance; they do so ideologically and metaphorically. (D. Strasser, personal communication, April 13, 2017)

To understand how women experience their bodies, identities, and selves as runners, I framed my poetic analysis using feminist standpoint theory because of the recognition that knowledge takes place in particular embodied contexts. "Listening to the body is as valuable and its physicality is as integral to our lives as listening to spiritual, philosophical or intellectual guidance" (Snowber, 2016, p. xiii). Our knowledge, including bodily knowledge, and our accounts of this knowledge are neither neutral nor universal; they are socially constructed (Collins, 1986; Harding, 2004). The body as it is lived and constructed through gender, sexuality, and other social positioning is important for feminist theorists to attend to and connect women's actual experiences in their bodies with the political project of deconstructing the mind–body split. "It [feminist analysis of the body] needs to explicitly tackle the relationship between the symbolic and the material, between representations of the body and embodiment as experience or social practice in concrete, cultural, and historical contexts" (Davis, 1997, p. 15). Embodied experience entails power and resistance. "We regulate ourselves, and we also practice resistance to regulation" (Ellingson, 2017, p. 89).

Examining embodiment for women runners means including bodily knowledge from their standpoints and perspectives to critique how dominant messages about women's bodies frame the experiences of women runners. The understandings of what it means to have a runner's body and to be a "real runner" may differ for those not situated as the recipients of privileged race, gender, and normative understandings (Collins, 1986). Being allowed to take up public space as a runner is also contingent on paradoxical expectations of exercising in public—women face potential harassment, attack, and threat as well as empowerment, agency, resistance, power, and pleasure (e.g., Allen-Collinson, 2010). When women shared their stories with me, I attended to how "the influence of circulating norms" shape "perceptions of appropriate choices for embodied performances of self" (Ellingson,

2017, p. 88). Some women resisted the normative scripts, some reinscribed them, and others altered the scripts through their insistence on taking up space as runners. Allen-Collinson (2010) studied women's running from a phenomenological perspective focusing attention on the lived experiences of running. "Battling the elements, active social and corporeal agency, resistance, and transformative action certainly constitute core elements in my own lived experience of training for distance running, and are also reflected in accounts of women's physical activity as resistance" (Allen-Collinson, 2010, p. 285). Women experienced running as social and solitary, pleasurable and painful, dangerous and empowering. I presented women's running experiences, highlighting their bodily experiences as connected to cultural practices through narrative, poetry, and poetic transcription.

Poetry as Running Logs

In my field notes, I found poems. I wrote many entries in my field notes as poems, and constructed many poems while running. When I think about what makes poetry, I typically think of the line "Poetry is the sound of language in lines … Line is what distinguishes our experience of poetry as poetry, rather than some other kind of writing" (Longenbach, 2008, p. xi). And when thinking of the line, I think of breath. For me, running is also about breath; breathing in the pleasure, breathing out the hurt. Breathing in who I want to be, breathing out lesser versions of self. Trying to breathe in through the nose and out through the mouth to catch my pace during a run. "The body is rooted in breath, rhythm, and poetry" (Snowber, 2016, p. xv). Thus, I used poetry as representation and analysis of women's running stories.

The haiku as running log that I interwove with my own running stories in Chapter 2 represents my embodied practice. I used haiku and other poetic forms as part of my field notes and research practice:

1. to show embodiment and to mirror the rhythm of running;
2. to demonstrate the process of analysis;
3. to connect writing and ethnographic practice.

First, the use of poetry is one way to show embodiment. The use of the lyric in poetry can have "a reader come away with the resonance of another's world" (Neilsen, 2008, p. 96). Haiku are lyric poems that "stress moments of subjective feeling and emotions in a short space" (Faulkner, 2017, p. 218). The Japanese form consists of 3 lines with 17 syllables in a 5/7/5 syllable line count. Traditionally, haiku were present-tense poems that focused on associative images and included a season word (i.e., *kigo*) and a focus on nature. Modern interpretations break many of these rules. "However, the philosophy of haiku has been preserved: the focus on a brief moment in time; a use of provocative, colorful images; an ability to be read in one breath; and a sense of sudden enlightenment and illumination"

116 Running as Feminist Embodiment

(poets.org, 2016, para. 7). The use of poems to help show what running feels like "becomes *embodied experience* when audiences feel *with*, rather than *about* a poem; they experience emotions and feelings *in situ*" (Faulkner, 2017, p. 226). The goal with using poems about running was to have you feel like you are running with me and other women, to feel our gasps and intake of breath, to breathe in rhythm with the words.

Second, I used poetry to demonstrate the process of analysis. Ethnographic poetry is a way to meditate on field notes (Kusserow, 2008). The poems that represented running logs, and the poems I wrote during field work, helped me to analyze women's stories of running on and offline. "Writing reflective poems helps researchers ask more focused questions, and questions that they may not have considered" (Faulkner, 2017, p. 214). Poetry from my fieldwork ended up being part of the analysis process. For example, I found the following poem in my field notes.

★★★

History of Body

Feet crack like chipped ice,
corrective shoes squeak all secrets.

The snap of elastic as middle-age spread
creeps across your health charts—still in

normal range of what can be measured,
the righteous rage of boxed in body

ricochets off political signs in neighbor's yards.
You run 13 miles, take a nap.

This body scabs history from mother
to daughter into the shower of flower petals

at the finish line, the medal of survival,
heavy and bold   around a young fresh neck

born running, the birth scars
shine in the light   run & scab.

★★★

I wrote, "History of Body," as I was musing on my participation in road races at a writing conference. The poem helped me see the importance of running as relationship, even for this ethnographer who studies and teaches about personal relationships.

Excerpts from my field notes demonstrate how some prose became poems. I wrote some of the poems in my head as I ran, such as the haiku. Other poems were created from passages in my field notes. For example, I wrote the following excerpt after a run in the country where I was chased by dogs mile after mile:

> I think of running as the embodiment and the irony that running is bodily and what holds one back is the mind. Thoughts running 6 miles- Mind/ Body split becomes MINDBODYMINDBODY. I want to kill the dogs as they run up on my back end, and I imagine how they will maul my calves, blood and running gristle as I tumble down the gravel road. There are different impediments/challenges as we run. In the country here, I need pepper spray for the dogs. And in the city, the pepper spray for errant pigeons and the occasional geese. And potential harassment from humans. (June, 2016)

This passage found its way into some haikus (e.g., Body is Mind) and a poem I wrote about the massacre of queer people of color at Pulse nightclub in Orlando, Florida while on vacation with my extended family.

★★★

1. Country Run

Which path do you choose?
A. The gravel road with the bucolic view
and the loose dogs who want,
hungry for a taste of your salt
and the sweet smell of fear
clinging to your running clothes,
and you, without a weapon,
no pepper spray or knife,
a stranger who only packed your wit
the dogs will maul like a chewy
piece of out-of-town gristle.

OR

B. The country highways paved with Capital
Letters, AA and AB, where you can count
the ripening road kill and play
the game of human frogger:
you keep a limb if you whip

118 Running as Feminist Embodiment

your head fast enough around the curves
to hop into the ditch moments before
the rattle of the truck whisks wind
in your hair and dirt
in your bulging eyes.

If you twist an ankle,
deduct 10 points. If you get
a wave, go back to KK Highway
and run another mile.

OR

C. Try the private roads
in gun country and work out riddles
while you practice fight and flight–
like if there are 300 million guns
in the US, how many reside on Private Drive?
How much of the stockpile of ammo
smelted and stocked in the side garage
would it take to stop you
and your strange running shoes
from trespassing where you don't belong?

★★★

This poem is an example of how I used poetry as analysis (e.g., Faulkner, 2014). Poetic analysis is a technique of using poems as data for qualitative analysis (Faulkner, 2009). I read the poetry in my field notes, the poems composed from field notes, and the poetic transcription of women's running stories to analyze the embodied experiences of running. I used poetic transcription of the interviews in Chapter 3, a method of representing participants' speaking styles and world-views through poetic lines (Madison, 2004). I took women's exact words and language from interviews and arranged them in lines that mirrored speaking styles and breath (e.g., Butler-Kisber, 2002), arranging their words to resemble their performances of the oral (Madison, 1991). I could make an argument that these transcriptions are like found poems (cf. Walsh, 2006), poems that come into being by paying attention to line, language, and sound (Padgett, 1987).

Poetic analysis also helped me consider the form this ethnography needed to be written and composed in, including the visual and sound elements. "The power of poetic inquiry can be realized if we ride the dialectic between aesthetic and epistemic concerns" (Faulkner, 2017, p. 221). The ethnography needed to be an aesthetic experience as well as an analytical one. And my story as runner and as feminist ethnographer needed to be interwoven given that "the ethnographer's writing self *cannot not* be present, there is *no* objective space outside the text" (Denzin, 2014, p. 26, emphasis in original).

Third, the use of poetry helped me to connect writing and ethnographic practice; what I think about on runs and how I write connected with poems as ligaments and tendons. I end with a passage from a poem about the 2016 Pulse nightclub massacre I wrote because I did what I do when I need to think, to process news, to work out my writing and emotions: Run. Run like a …

***

Caution: Black Bears

I do what I do when I can't think—
run in the woods alone
though the sign cautions
about the Black Bears and
we all know what women alone
are asking for and
my shoulder is jacked and
the path is all rocks
to be tripped over but
at least my orthotics squeak and
I stink so maybe
that will scare them away and
who wants to eat
a bisexual feminist who needs
"to lighten up and take a joke" and
who is "dressed like this" and
who wonders if her family would care
if they couldn't find her and
I could have been in that nightclub and
I like bars and clubs where women
look at women with love and
it could have been me or
my girlfriend in that nightclub but
it was all of us, really, and
why can't they see that and
I have to pee so
if I pee here
will that make it more likely
I will be mauled or
will that mark this spot as mine and

those brothers and sisters and
sons and daughters were just dancing and
flirting and being all joy and
I need to run my way out of sorry
and to the thought
our responses are different but
we are all afraid we are all
afraid and we are all
in this we are all
in this. We are all.

References

Allen-Collinson, J. (2010). Running embodiment, power and vulnerability: Notes towards a feminist phenomenology of female running. In E. Kennedy and P. Markula (Eds.), *Women and exercise: The body, health and consumerism* (pp. 280–298). London: Routledge.

Bernstein, E. E., & McNally, R. J. (2015). Acute aerobic exercise helps overcome emotion regulation deficits. *Cognition and Emotion, 31*(4), 834–843. doi:dx.doi.org/10.1080/02699931.2016.1168284

Butler-Kisber, L. (2002). Artful portrayals in qualitative inquiry: The road to found poetry and beyond. *Alberta Journal of Educational Research, XLVIII*(3), 229–239.

Collins, P. H. (1986). Learning from the outsider within: The sociological significance of Black feminist thought. *Social Problems, 33*(6), S14–S32. http://ucpressjournals.com/journal.asp?jsp

Collins, P. H. (2000). *Black feminist thought: Knowledge, consciousness, and the politics of empowerment.* New York, NY: Routledge.

Creer, D. J., Romberg, C., Saksida, L. M., van Praag, H., & Bussey, T. J. (2010). Running enhances spatial pattern separation in mice. *Proceedings of the National Academy of Sciences of the United States, 107*(5), 2367–2372. doi:10.1073/pnas.091172510

Davis, K. (1997). Embodying theory: Beyond modernist and postmodernist readings of the body. In K. Davis (Ed.), *Embodied practices: Feminist perspectives on the body* (pp. 1–23). London: Sage.

Denzin, N. K. (2014). *Interpretive autoethnography* (2nd ed.). Thousand Oaks, CA: Sage.

Ellingson, L. (2017). *Embodiment in qualitative research.* New York, NY: Routledge.

Faulkner, S. L. (2009). *Poetry as method: Reporting research through verse.* New York, NY: Routledge.

Faulkner, S. L. (2014). *Family stories, poetry, and women's work: Knit four, frog one.* Rotterdam: Sense.

Faulkner, S. L. (2016). Writing the personal: How to begin and maintain a writing practice. "Writing from the heart" panel presentation at the 12th International Congress of Qualitative Inquiry, Urbana-Champaign, IL.

Faulkner, S. L. (2017a). Writing practice: A narrative poem. *International Review of Qualitative Research, 10(3),* 238–241.

Faulkner, S. L. (2017b). Poetic inquiry: Poetry as/in/for social research. In P. Leavy (Ed.), *The handbook of arts-based research* (pp. 208–230). New York, NY: Guilford Press.

González, M. C. (2000). The four seasons of ethnography: A creation-centered ontology for ethnography. *International Journal of Intercultural Relations, 24*(5), 623–650. doi:10.1016/S0147-1767(00)00020-1

Harding, S. (2004). Introduction: Standpoint theory as a site of political, philosophic, and scientific debate. In S. Harding (Ed.), *The feminist standpoint theory reader: Intellectual and political controversies* (pp. 1–15). New York, NY: Routledge.

Kusserow, A. (2008). Ethnographic poetry. In M. Cahnmann-Taylor & R. Siegesmund (Eds.), *Arts-based research in education: Foundations for practice* (pp. 72–78). New York, NY: Routledge.

Lee, Y. (2005, October 3). I'm a runner: Haruki Murakami. This novelist uses his running to make his books top-notch. *Runner's World*. Retrieved from www.runnersworld.com/celebrity-runners/im-a-runner-haruki-murakami

Longenbach, J. (2008). *The art of the poetic line*. Saint Paul, MN: Graywolf Press.

McCaughey, M. (1997). *Real knockouts: The physical feminism of women's self-defence*. New York, NY: New York City Press.

Madison, D. S. (1991). "That was my occupation": Oral narrative, performance, and black feminist thought. *Text and Performance Quarterly, 13*, 213–232.

Madison, D. S. (2004). Performance, personal narratives, and the politics of possibility. In Y. S. Lincoln & N. K. Denzin (Eds.), *Turning points in qualitative research: Tying knots in a handkerchief* (pp. 469–486). Walnut Creek, CA: AltaMira.

Menzies-Pike, C. (2017). *The long run: A memoir of loss and life in motion*. New York, NY: Crown.

Moraga, C., & Anzaldúa, G. (Eds.). (1981). *This bridge called my back: Writings by radical women of color*. New York, NY: Kitchen Table.

Murakami, H. (2008). *What I talk about when I talk about running*. New York, NY: Knopf.

Neilsen, L. (2008). Lyric inquiry. In J. G. Knowles & A. L. Cole (Eds.), *Handbook of the arts in qualitative research: Perspectives, methodologies, examples, and issues* (pp. 93–102). Thousand Oaks, CA: Sage.

Oates, J. C. (2003). *The faith of a writer: Life, craft, art*. New York, NY: Harper Collins.

Padgett, R. (Ed.). (1987). *The teachers and writers handbook of poetic forms*. New York, NY: Teachers and Writers Collaborative.

Poets & Writers (2016, July 13). The thing and the other thing. *Poets & Writers*. www.pw.org/content/the_thing_and_the_other_thing

poets.org. (2016, March 1). Haiku: Poetic form. poets.org. Retrieved from www.poets.org/poetsorg/text/haiku-poetic-form

Presland, C. (2016, January 13). Three ways that running boosts your creativity. Author Unlimited. Retrieved from http://authorunlimited.com/running-and-creativity/

Snowber, C. (2016). *Embodied inquiry: Writing, living and being through the body*. Rotterdam: Sense.

Toor, R. (2014, June 23). What writing and running have in common. *The Chronicle of Higher Education*. Retrieved from http://chronicle.com/article/What-WritingRunning-Have/147193/

Velija, P., Mierzwinski, M., & Fortune, L. (2013). "It made me feel powerful": Women's gendered embodiment and physical empowerment in the martial arts. *Leisure Studies, 32*(5), 524–541. doi:10.1080/02614367.2012.696128

Walsh, S. (2006). An Irigarayan framework and resymbolization in an arts-informed research process. *Qualitative Inquiry, 12*(5), 976–993.

APPENDIX

Chapter 3: Women Who Run

Interview Questions/Protocol

I am interested in women's embodied experiences with running: how they run, how running fits into the context of their lives and relationships, how they enact or challenge cultural scripts of women's activities and normative running bodies, and what running means for their lives and identities.

This project was approved by BGSU Institutional Review Board (IRB) May 7, 2014.

Questions about Identity and Identity as a Runner

1. I am _____. Fill in the blank with relevant identities.
   If runner is not on the list, ask why? What makes one a runner?
   Basic demographic info: Age? Ethnic background? Sexual Orientation? Gender Identity? Relationship status? Education level? Work/Job/Career? Children?
2. Please describe what this picture/image means to you. (Participant will have selected a photo/image that reflects their experiences running).
3. What is it about the image that you like? How does it depict your experiences running?

I Want to Ask Questions about the (Embodied) Meaning of Running in Your Life

1. Tell me how you began to run.
2. How would you describe your running body? What are your strengths? What do you like about your running body?

Appendix **123**

3. What does running mean to you? For your daily life? For your future?
    3a. What does it mean to be a woman who runs?
4. How does it feel to run (before, during, after)? Can you describe a good run? Can you describe a bad run? (Do you like your surroundings when you run? e.g., catcalls.)
5. What do you like most about running? What do you like least about running?
    5a. How would you describe your running?

How Do You Run? I Want to Ask Questions about the Specifics of Your Running

1. When do you run? How often? Where do you run? Describe a typical week in your running life.
2. Do you use any online running coaches or running apps that help you train/ track your running (e.g., run coach, daily mile)? If yes, why do you use them? If no, do you keep track of your running? Do you have some system for training? Do you participate in races? If yes, what do you like about races? If no, what keeps you running?
3. Have you ever been injured because of running? Please describe the injury. What led to it? How did it impact your running? Any pain issues?
    3a. Nutrition and running.
4. Do you run in races? What do you like about races? How many races do you do a year? Do you know how many races you have participated in?
5. Do you ever look up information online about running? Do you subscribe to any running magazines? If yes, what do you think of the information?
6. What do you think of portrayals of women runners in the media (e.g., online, magazines)?

Running and Your Social Network

1. Do you run with others? Who? Why do you run with others? In what context do you run (e.g., run for charity)?
2. How do others talk about your running? Your running body? Do you have running friends?
3. How do significant others support or not support your running?
4. What role (if any) do others play in your running?

Demographic Information

Interviews with women runners (N = 41) in 2014

> **Age** M = 37 years old; range = 25–56 years old
> **Education** high school = 4; some college = 3; associate = 2; BS = 1; BA = 12; MA = 11; MFA = 1; PhD = 7; PhD candidate = 4

Ethnicity Latinx = 2; Japanese American = 1; White/African American = 1; White/Bulgarian = 1; White = 36

Locale in the United States: AK = 1; CA = 1; CO = 2; FL = 2; GA = 1; IL = 4; IN = 1; KS = 2; LA = 1; MA = 1; MN = 1; MO = 1; MS = 1; NJ = 1; NY = 2; OH = 4; PA = 1; UT = 6; VA = 3; WA = 5

Children zero = 21; deceased = 1; miscarriage = 1; one = 4; two = 12; three = 2; four = 1

Sexual orientation lesbian/queer = 4; heterosexual = 37

Relational status single = 4; married = 28; divorced = 8; engaged = 2; dating = 1

Occupation professor = 6; nurse = 3; student = 8; teacher = 2; copywriter; executive assistant; project manager; environmental education coordinator 4H; working professional = 3; personal trainer; biologist; nutrition coach; retired; senior marketing coordinator; assistant director Library; supervisor Black Jack Casino; communication specialist; army wife; spin instructor; SAHM; sales = 2; assistant coach cross-country team; accountant; caterer

Chapter 4: Method

In Faulkner, Tetteh, and Behrmann (2016), we used qualitative content analysis to examine two popular running sites that focused on women's running (Krippendorff, 2004; Mayring, 2000). We selected *Zelle* (September 2014 to December 2014) and *Women's Running* (September 2014 to December 2014) due to their prominence in the running industry and the high volume of traffic they enjoy. We were interested in how women runners experienced their running selves through an examination of their own stories, so qualitative content coding was important in determining latent themes concerning identity (e.g., real runners) that are not easily quantifiable (Schreier, 2012). We used Carpenter's (2002) distinction between manifest and latent content to define a difference between the general content of women who run in conjunction with their identity as runners and underlying themes that indicate stances toward identity and embodied experience (e.g., how stories reinforced or subverted dominant cultural understandings of women runners).

Analysis

We used four separate coding meetings to establish our topics and focus of analysis in the project. In the first meeting, we looked together at visual images, text, and comments on popular running sites and blogs targeted to women. We asked a series of questions: How are women who run portrayed? How are the stories of women who run portrayed? What stories are portrayed? How is empowerment defined in these spaces? How do the images portrayed reflect class, race, ableism, size, gender, presentation, gender orientation, and motherhood? How does image

Appendix **125**

define what a runner is and should be? After discussion, we set up a second meeting to talk about what themes seemed most prevalent in the running sites and on what we should focus our analysis. We agreed to examine women who are not elite or paid athletes given our interest in the experiences of everyday women who run.

The third meeting was used to narrow down the online sources for analysis and the specific focus of our analysis. We decided that blogs focused on one woman's story were not sufficient for our purposes, thus we selected the two most popular sites targeted toward women who run, *Zelle* (from *Runner's World*) and *Women's Running*. We each selected the same articles from the sites to analyze for the next meeting. Specifically, we decided to examine articles and subsections of articles, reviews, columns, and forums that concerned women who run. We included visual and textual information.

In the fourth meeting, we determined to analyze *Zelle* from its inception through December 2014 and, for comparison purposes, to analyze *Women's Running* during the same time frame. The specific things we focused on included the articles and comment sections within the website tabs training and inspiration, nutrition and travel, health and wellness, and blogs. We were interested in *appearance* (outfits, physical appearance, traditional body, post-baby bodies); *relationships* (mothers who run); *female empowerment* (me time, run like a girl, inspirational stories); and *workout tips/guidelines* (nutrition, tips).

Each author selected one theme to analyze: 1) identity (what does it mean to be a woman runner); 2) appearance (nutrition, health, and looks); 3) relationships (what does it mean to be a mother and a runner?). We used the following questions to guide our analyses: How are women who run portrayed? What does it mean to be a woman runner? How is information about women runners portrayed (personal, research)?

References

Carpenter, L. M. (2002). Analyzing textual material. In M. W. Wiederman & B. E. Whitley Jr. (Eds.), *Handbook for conducting research on human sexuality* (pp. 327–343). Mahwah, NJ: Erlbaum.

Faulkner, S. L., Tettah, D. A., & Behrmann, E. M. (2016). Women who run: Identity, embodied experience, and gendered expectations in popular women's running websites. Paper presentation at the annual Central States Communication Association. Grand Rapids, MI.

Krippendorff, K. (2004). *Content analysis: An introduction to its methodology* (2nd ed.). Thousand Oaks, CA: Sage.

Mayring, P. (2000). Qualitative content analysis. *Forum: Qualitative Social Research, 1*(2), Art. 20. Retrieved from www.qualitative-research.net/fqs-texte/2-00/2-00mayring-e.htm

Schreier, M. (2012). *Qualitative content analysis in practice*. Thousand Oaks, CA: Sage.

INDEX

Allen-Collinson, J. 5, 7
athlete identity 21, 38

bad runs 61, 81–84, 108
bisexual 30–31, 39, 48, 119
Boston Marathon 2
Boy Scout Half Marathon 33, 54–55, 57

Catcalls 24, 74–75

definitions of runners 70, 77, 85–86, 96, 101, 114
dissertation 20–21, 23–24, 30

embodiment: embodied experience 4–5, 11, 19, 114–115; embodied practice 4; feminist embodiment 5, 13, 19, 33, 101, 108, 113–114; women's experiences of 5, 19, 91–92, 108

Faulkner, S. 4, 6, 9, 10, 12, 43, 91, 105, 107–108, 112–113, 115–116, 118
Fat Girl Running 1, 7, 90, 92, 95, 98
feminist ethnography 4, 7, 111–113, 119
feminist identity 13–15, 17, 20–21, 24, 27
feminist runner 6, 21
Flanagan, S. 2
Flying Pig Half Marathon 4, 32–33, 79

Gay Games IX 4, 6–7, 33, 37–50
good runs 61, 81–84, 108
Goucher, K. 5, 95, 112

haiku as running log 6–7, 10–20, 22, 24, 26–35, 37, 40–41, 43–44, 50, 53–57, 107, 115–117

lesbian identity 24–26, 30–31, 40, 42

male privilege 26
Marine Corps Marathon 2, 65, 69, 80
marketing of running 3, 5, 90–91, 102
marriage 20, 23, 27–28, 34–37, 53–54
media messages about women runners 1, 70, 93–95, 97, 101
Menzies-Pike, C. 1, 4, 5, 61, 66, 73, 78, 81, 87, 111, 112
middle-aged 9, 11, 29, 32, 38, 51, 55, 67, 110
mind-body connection 5, 56–57, 59, 66, 84, 92, 101, 114, 117
misogyny 17, 19, 24, 31, 61, 74–75
mother runner 3, 28–29, 32, 36, 49, 71, 100–101
Munich Marathon 4, 52–53

Oates, J. C. 105, 108–109
Olympics 2, 40

participant observation 4, 6, 25, 38, 113
personal record (PR) 44, 52–55, 70, 78, 84, 96
physical feminism 7, 11, 101, 113–114
poetic inquiry 4–5, 7, 64, 111–113, 115–119

Index **127**

queer identity 7, 31, 61, 78, 92, 112, 117, 124

reasons for running 61–64, 87–88
road races 76–78; 5k 1, 3–4, 10, 19–20, 38, 44–45, 64, 71, 100; 10k 1, 3–4, 26, 33, 38, 40–45, 49–50, 52–53, 67–68; half marathon 1–4, 21–22, 32, 44–45, 47, 52, 54–57, 66, 68, 76–77, 79, 107, 110; marathon 1–2, 26, 44–45, 47, 52, 55, 60, 62, 64–66, 68–69, 71, 73, 75–78, 80–81, 83–84, 88, 92–94, 108, 110; women only 3
runner-runner/real runner 20, 26, 39, 42, 61, 77, 92, 114, 124
Runner's World 1, 7, 20, 91–92, 111
running app 92–93, 110
running as: accountability 7, 69–70; effort 12, 17; expansion 21–22, 84–85; feminist practice 6, 11, 19, 33, 101, 108; for a reason 75–76; friendship 28, 31–32, 110–111; health 4, 7, 17, 65–69; identity 21, 27, 88, 95–96, 110–111; pleasure and danger 56; relational practice 4, 6–7, 11, 41, 64, 68, 70–73, 76, 116–117; self-care 22; safety and danger 5, 7, 14, 19, 73–75; self-definition 85–88, 101, 108; stress-release 25, 34; therapy 68–69

running blogs 1, 4, 7, 90, 92, 94–95, 98, 101, 124–125
running bodies 4–7, 27, 78–81, 91, 97–99, 111–112
running discourse 4–5, 7
running injuries 6, 9, 10, 12–13, 19, 21–23, 29, 33, 51, 55–57, 61, 66–67, 105
running practice 20–21, 54–55, 73, 82
running stories 6, 20–21, 60, 91, 114–115

sexual orientation 24
socio-cultural influences on running 4–5, 7, 53, 91–92, 94, 98–99, 101, 114, 115
standpoint theory 91–92, 113–115
Switzer, K. 1–4, 94

Title IX 2
training for races 27–28, 76–78, 110–111

violence against women 13, 16, 18–19, 24

Winfrey, O. 2
woman runner 1–2, 10, 30, 60–64, 86
Women's Running Magazine 1, 7, 90–92, 95, 98–99, 101, 124–125
writing and running 10, 105–112, 119

Zelle 7, 74, 91–92, 94–95, 98–99, 101, 124–125

Taylor & Francis eBooks

Helping you to choose the right eBooks for your Library

Add Routledge titles to your library's digital collection today. Taylor and Francis ebooks contains over 50,000 titles in the Humanities, Social Sciences, Behavioural Sciences, Built Environment and Law.

Choose from a range of subject packages or create your own!

Benefits for you
- Free MARC records
- COUNTER-compliant usage statistics
- Flexible purchase and pricing options
- All titles DRM-free.

REQUEST YOUR FREE INSTITUTIONAL TRIAL TODAY

Free Trials Available
We offer free trials to qualifying academic, corporate and government customers.

Benefits for your user
- Off-site, anytime access via Athens or referring URL
- Print or copy pages or chapters
- Full content search
- Bookmark, highlight and annotate text
- Access to thousands of pages of quality research at the click of a button.

eCollections – Choose from over 30 subject eCollections, including:

Archaeology	Language Learning
Architecture	Law
Asian Studies	Literature
Business & Management	Media & Communication
Classical Studies	Middle East Studies
Construction	Music
Creative & Media Arts	Philosophy
Criminology & Criminal Justice	Planning
Economics	Politics
Education	Psychology & Mental Health
Energy	Religion
Engineering	Security
English Language & Linguistics	Social Work
Environment & Sustainability	Sociology
Geography	Sport
Health Studies	Theatre & Performance
History	Tourism, Hospitality & Events

For more information, pricing enquiries or to order a free trial, please contact your local sales team:
www.tandfebooks.com/page/sales

The home of Routledge books

www.tandfebooks.com